Parallel Finite Volume Computation on General Meshes

Yuri Vassilevski · Kirill Terekhov ·
Kirill Nikitin · Ivan Kapyrin

Parallel Finite Volume Computation on General Meshes

Springer

Yuri Vassilevski
Marchuk Institute of Numerical Mathematics
of the Russian Academy of Sciences and
Moscow Institute of Physics and Technology
and Sechenov University
Moscow, Russia

Kirill Nikitin
Marchuk Institute of Numerical Mathematics
of the Russian Academy of Sciences and
Moscow State University
Moscow, Russia

Kirill Terekhov
Marchuk Institute of Numerical Mathematics
of the Russian Academy of Sciences
Moscow, Russia

Ivan Kapyrin
Marchuk Institute of Numerical Mathematics
and Nuclear Safety Institute of the Russian
Academy of Sciences
Moscow, Russia

ISBN 978-3-030-47231-3 ISBN 978-3-030-47232-0 (eBook)
https://doi.org/10.1007/978-3-030-47232-0

This Springer imprint is published by the registered company Springer Nature Switzerland AG
The registered company address is: Gewerbestrasse 11, 6330 Cham, Switzerland

Acknowledgements

This book would not appear without dedications of many colleagues and collaborators. The coauthors of our papers cited in the bibliography are our collaborators to whom we are greatly indebted. In particular, we wish to thank Konstantin Lipnikov, Daniil Svyatskiy, Alexander Danilov, Alexey Chernyshenko, Konstantin Novikov, Vasiliy Kramarenko, Ruslan Yanbarisov, Mikhail Shashkov, Bradley Mallison, Hamdi Tchelepi, and Maxim Olshanskii for their contribution to our joint papers and cooperative research on the nonlinear FV methods. We are also grateful to Mary Wheeler, Serguei Maliassov, Ilya Mishev, Roland Masson, and Denis Voskov for illuminating discussions on applications of the FV methods in reservoir simulation; Jerome Jaffre and Alexander Rastorguev for introduction to simulation of radionuclides subsurface migration; Igor Linge and Sergey Utkin for practical RW disposal safety assessment problem formulations and methodology; Vitaly Volpert and Anass Bouchnita for joint development of the multi-physics model of blood flow coagulation; Denis Anuprienko, Fedor Grigorev, Georgiy Neuvazhaev, Viktor Suskin, and all the GeRa code developers for the cooperation in groundwater flow and radionuclides transport models development and their verification. We owe to our colleagues Igor Kaporin, Igor Konshin, Vadim Chugunov, Sergey Goreinov, and Sergey Kharchenko for fruitful discussions on incomplete LU factorization methods and contribution to INMOST software. We are in a great debt to Yuri Kuznetsov who introduced us to the world of mathematical modeling years ago.

A large part of the work presented here was supported by the ExxonMobil Corporation within a long-term research project at the Marchuk Institute of Numerical Mathematics of the Russian Academy of Sciences (INM RAS). We acknowledge the financial support by the Russian Science Foundation projects 18-71-10111, 19-71-10094, and the Russian Foundation for Basic Research projects 18-31-20048 and 19-31-90110, the RAS Research program 26 “Basics of algorithms and software for high performance computing” and the world-class research center “Moscow Center for Fundamental and Applied Mathematics”. Nuclear Safety Institute of the Russian Academy of Sciences (IBRAE) provided resources for the GeRa code development. We thank INM RAS and IBRAE, for administrative support of our research within the above projects.

Finally, we would like to thank all Springer team for making this publication possible. And, most importantly, we are grateful to our families for the patience and irretrievably lost time, otherwise spent by the authors writing the code and text of the book.

Contents

Acronyms

DDF	Density-driven flow
DFN	Discrete fracture network
DGR	Deep geological repository
DMP	Discrete maximum principle
DSA	Direct substitution approach
EBS	Engineered barriers system
EDFM	Embedded discrete fracture method
FV	Finite volume
FVM	Finite volume method
HLW	High-level radioactive waste
IMPES	Implicit pressure–explicit saturation
LILW	Low- and intermediate-level radioactive waste
mEDFM	Monotone embedded discrete fracture method
MFV	Monotone finite volume method
MPFA	Multi-point flux approximation
NMPFA	Nonlinear multi-point flux approximation
NTPFA	Nonlinear two-point flux approximation
NWC	Near-well correction
PDE	Partial differential equation
RHS	Right-hand side
RW	Radioactive waste
SIA	Sequential iterative approach
SNIA	Sequential non-iterative approach
TPFA	Two-point flux approximation

Notations for the Book

Mesh Elements

entity	as element	as vector
vertex	v, v_i	$\mathbf{v}, \mathbf{v}_i$
edge	e, e_{ij}	$\mathbf{e}, \mathbf{e}_{ij} \equiv \mathbf{v}_j - \mathbf{v}_i$
face	$f, f_{i\dots k}$	$\mathbf{n}_f$, $\mathbf{n}_{i\dots k} \equiv \mathbf{e}_{ij} \times \mathbf{e}_{ik}$

entity	as element	via vertices
cell	ω	$\omega(v_{i_1}, \dots, v_{i_k}) = \omega(i_1, \dots, i_k)$
v-superelement	$\sigma(v)$	$\bigcup_{\omega_i \ni v} \omega_i$
e-superelement	$\sigma(e)$	$\bigcup_{v_i \in e} \sigma(v_i)$
ω-superelement	$\sigma(\omega)$	$\bigcup_{v_i \in \omega} \sigma(v_i)$

Domain

mesh	Ω_h
domain	Ω
boundary	$\partial\Omega$
interface or a part of boundary	Γ

Vector Quantities

coordinate	$\mathbf{x} = (x_1, x_2, x_3)^T$ or $\mathbf{x} = (x, y, z)^T$
velocity	$\mathbf{u} = (u_1, u_2, u_3)^T$ or $\mathbf{u} = (u, v, w)^T$
auxiliary velocity vector	$\mathbf{w}$
normal vector	$\mathbf{n}$
external forces	$\mathbf{f}$
velocity boundary data	$\mathbf{u}_{in}$ or $\mathbf{u}_D$ or ...
vector flux	$\mathbf{q}$
gradient	∇

Tensor Quantities

Cauchy stress tensor	σ
diffusion or permeability tensor	$\mathbb{K}$
diffusion-dispersion tensor	$\mathbb{D}$

Scalar Quantities

pressure	p
collocated pressure	P_i
concentration	c
collocated concentration	C_i
density	ρ
Lame constants	μ, λ

Other Quantities and Conventions

spatial mesh size	h
time step	Δt
time instances	$t_0, t_1, \ldots$
Differential operators	$\mathrm{div}\,, \mathrm{rot}\,, \Delta$
Partial derivatives	$\frac{\partial f}{\partial t}$ or $\frac{\partial f}{\partial x}$
Scalar product for vectors	$\mathbf{x} \cdot \mathbf{y}$
Direct product for tensors	$\mathbf{X} : \mathbf{Y}$ ($\mathbf{X} : \mathbf{Y} = \mathrm{tr}(\mathbf{X}\mathbf{Y}^T)$)
Norms and scalar products	$\|\cdot\|$ and $(\cdot,\cdot)$ for L^2, while other norms and products are clearly labeled
Sobolev and Lebesgue spaces	$H^m(\Omega)$, $H^m(\Omega)^d$, $L^p(\Omega)$
dependence on time t_k	$f^k = f(t_k)$ (use upper indexes)
eigenvalues	$\lambda_1, \lambda_2, \lambda_3$

Graph Structure

set of vertices	$\mathcal{V}$
set of edges	$\mathcal{E}$
graph	$T = (\mathcal{V}, \mathcal{E})$
vertex	v
vertex coordinates	$\mathbf{x}(\mathrm{v})$
edge	$\mathrm{e} = (\mathrm{v}_1, \mathrm{v}_2)$
neighborhood vertices	$\mathcal{N}(\mathrm{v})$
neighborhood edges	$\mathcal{E}(\mathrm{v})$

Chapter 1
Introduction

1.1 Rationale

The book presents in a systematic way a methodology for the development of parallel multi-physics models and its employment in geophysical and biomedical applications. The methodology includes conservative methods of discretization of PDEs on general meshes, as well as data structures and algorithms for organization of parallel simulations on general meshes. The structures and algorithms form the core of the Integrated Numerical Modeling Object-oriented Supercomputing Technologies (INMOST) platform for development of parallel models on general meshes. For applications, we consider geophysical and biomedical challenges. The geophysical applications address radioactive contaminant propagation with subsurface waters and reservoir simulation. The biomedical application deals with clot formation in blood flow.

1.2 Objectives

Several important features distinguish this text from other monographs on parallel computing. First, we stay on a solid ground of conservative nonlinear finite volume discretizations on general meshes enjoying monotonicity of the discrete solution, the second-order accuracy, and the compact discretization stencil. The latter is crucial for minimization of interprocessor communications. Second, the book introduces to English-speaking world our toolkit for parallel models development, the INMOST platform, www.inmost.org. INMOST is a tool for supercomputer simulations characterized by a maximum generality of supported computational meshes, distributed data structure flexibility, and cost-effectiveness, as well as cross-platform portability. Third, we describe algorithms for parallel operations with data on general meshes. Fourth, we present our approach to the solution of nonlinear systems arising after discretization of multi-physics problems. The approach is based on automatic

Y. Vassilevski et al., *Parallel Finite Volume Computation on General Meshes*,
https://doi.org/10.1007/978-3-030-47232-0_1

differentiation. Fifth, we address topical geophysical and biomedical challenges. Summarizing, the main thrust of the book is to provide in one text all ingredients of our methodology, from algorithms and numerical methods to the open-source software, and examples of practical applications. Therefore, applied mathematicians, computer scientists, and engineers may find this monograph to be a useful resource.

1.3 Structure and Overview of the Book

The structure of the book is oriented to a reader interested in our methodology. His interest may be caused by the need of general meshes use, or the need of fast and efficient parallel model development, or the need of modeling complex phenomena with multiple stiff physical processes, or all together.

Since the methodology is based on numerical methods, we first introduce the finite volume (FV) method on general meshes. In Chap. 2, we define general polyhedral meshes forming a consistent tessellation of the computational domain and present the cell-centered FV method on such meshes. The advantage of the method is its explicit conservativity which implies a cellwise balance principle: a source on a cell is counterbalanced by fluxes through the cell boundary. The wide diversity of FV methods [58, 65] is provided by different approaches to flux approximation (FA).

In our methodology, we adopt such flux approximations that produce monotone schemes. Monotonicity of a numerical scheme may imply either solution positivity or the discrete maximum principle. Positivity is important for concentrations, density, energy, absolute temperatures, etc. The discrete maximum principle guarantees the absence of artificial extrema in the numerical solution; for numerical pressure (head), this implies the absence of non-physical Darcy flows from cells with lower pressure to cells with higher pressure.

A monotone scheme can be built using a simple linear two-point flux discretization which, however, does not provide flux approximation in the general case of polyhedral meshes and tensor diffusion/permeability coefficients. Nonlinear flux discretizations generate monotone FV schemes at the cost of scheme nonlinearity, even if it is applied to a linear PDE. In Chap. 2, we present monotone FV methods with flux discretizations contributing to our general methodology and analyze their monotone properties in application to the diffusion and convection–diffusion equations. Also, we introduce a new concept of linear two-point vector flux discretizations applicable to systems of partial differential equations. Being linear discretizations, they demonstrate higher than the first-order approximation (but less than the second-order, in general) and do not provide theoretically monotone properties. However, the concept is appealing in multi-physics applications as the vector flux discretizations are monotone (according to numerical evidence) and are stable in spite of degrees of freedom collocated in cell centers.

Chapters 3 and 4 present applications of the considered FV methods to the approximate solution of two industrial challenges, reservoir simulation and assessment of radioactive waste disposal facilities' safety. The black-oil reservoir model involves simultaneous solution of three Darcy laws that describe the mixture of water, oil, and gas. Modeling radionuclide migration in the geosphere involves the solution of flow and transport problems. Here, we consider single-phase flow with possibly variable saturation and active buoyant forces. Transport model includes basic advection–diffusion–dispersion mechanisms as well as chemical reactions, radioactive decay, and sorption. Both applications involve multi-physics phenomena, and both require development of a computational beginning-to-end technology. Another common feature of these applications is adoption of the monotone FV schemes in elementary differential operators (diffusion and convection) separately, based on the schemes for the convection–diffusion equations.

Chapter 5 addresses blood coagulation processes. The blood coagulation model couples the Navier–Stokes equations with a Darcy term and nine additional advection–diffusion–reaction equations that participate in reactive cascade during coagulation of the blood. The chapter presents another approach to FV discretization of complex phenomena with multiple stiff physical processes. In this approach, the flux discretization is derived for two differential equations simultaneously. This provides a very robust and stable numerical scheme which grants large time steps even in the presence of blood coagulation reactions.

Development of parallel models of complex phenomena puts a significant burden on the programmer. He has to implement numerical methods, manage the unstructured grid and data exchanges with MPI, assemble large distributed algebraic systems, solve the resulting linear and nonlinear systems, and finally postprocess the result. INMOST platform alleviates most of the complexity and provides a unified set of tools to address each of the aforementioned issues. In Chap. 6, we present these tools to the reader. We consequently address managing data structures and operating with mesh data, solution of systems of linear and nonlinear algebraic equations, and parallel multi-physics model development.

The book introduces INMOST, but INMOST is not the only toolkit for multi-physics modeling: one may enjoy commercial alternatives COMSOL [2], ANSYS Fluent [1], Star-CD [13] and open-source alternatives Dumux [66], OPM [11], Elmer [5], OOFEM [9], OpenFOAM [10], SU2 [14], CoolFluid [3], and many others. A comparison of some of these packages is available in [27]. General perspectives for multi-physics software are discussed in [94].

Important feature of all these packages is that they provide modeling environment with an integrated set of computational methods which have limitations on physics, time-stepping, and couplings. For instance, our attempt to apply OpenFOAM package to blood coagulation simulation in reasonable computing time failed: provided methods do not support the fully coupled approach.

In contrast, INMOST does not contain integrated computational methods but provides a programming platform to implement them. Among other programming platforms, Dune [4], Trilinos [15], and PETSc [12] are worth mentioning. Dune provides tools for distributed mesh management. Trilinos and PETSc are widely used for built-in parallel linear and nonlinear solvers and seamless integration of third-party linear solvers but rely on third-party libraries for mesh management. Trilinos provides Sacado package for automatic differentiation. The framework for multi-physics simulations is under active development in Dune (within Dumux project [66]) and in Trilinos (within Amanzi project [49]). Functionality of these platforms can be used in large extent to build a simulator. INMOST platform enjoys the integration of Trilinos and PETSc tools.

In addition, there are a number of C/C++ libraries that solve a particular task. For mesh management, the incomplete list contains MSTK, MOAB, libMesh, and FMDB. For assembly and solution of linear systems, we mention Trilinos, PETSc, SuperLU, MUMPS, Hypre, and many others. For automatic differentiation, Sacado (Trilinos), ADOL-C, FAD, ADEL, and Adept have been developed. For nonlinear solvers, one can use Trilinos, PETSc, SUNDIALS, Ipopt, Snopt, and so on. The advantage of using separate tools stems from a greater level of maturity of popular libraries; the disadvantage is the absence of tight integration that is needed inevitably for construction of multi-physics framework.

Chapter 2
Monotone Finite Volume Method on General Meshes

Cell-centered finite volume (FV) discretizations are appealing for the approximate solution of boundary value problems since they are locally conservative and applicable to general meshes, i.e., to meshes with general polyhedral cells. In this chapter, we introduce nonlinear flux discretizations which result in monotone FV schemes at the cost of scheme nonlinearity, even if it is applied to a linear partial differential equation (PDE) such as diffusion and convection-diffusion equations. Also, we give two examples of linear two-point flux vector discretization of the diffusion equation in the mixed formulation and the Navier-Stokes equations. Such flux vector discretizations are stable in spite of degrees of freedom collocated at cell centers, are applicable to systems of PDEs, and demonstrate monotone numerical solutions.

2.1 Cell-Centered Finite Volume Method on General Meshes

The book deals with general polyhedral meshes, consistent tessellations of a given polyhedral domain Ω into a set C of polyhedral cells ω with planar faces f:

$$\Omega_h = \bigcup_{\omega \in C(\Omega_h)} \omega.$$

We assume that each cell is a star-shaped 3D domain with respect to its barycenter, and each face is a star-shaped 2D domain with respect to face barycenter. Otherwise, collocation of future degrees of freedom should be shifted from the barycenters. We denote by $\mathcal{F}_I, \mathcal{F}_B$ disjoint sets of interior and boundary faces of Ω_h, respectively.

We start the introduction to nonlinear finite volume discretizations from the diffusion problem in Ω for a scalar unknown $p \in H^1(\Omega)$:

$$\begin{aligned} -\mathrm{div}\,(\mathbb{K}\nabla p) &= g, \text{ in } \Omega, \\ \alpha p + \beta \mathbf{n} \cdot (\mathbb{K}\nabla p) &= \gamma, \text{ on } \partial\Omega, \end{aligned} \tag{2.1}$$

Y. Vassilevski et al., *Parallel Finite Volume Computation on General Meshes*,
https://doi.org/10.1007/978-3-030-47232-0_2

where $\mathbb{K}$ is piecewise-constant symmetric positive-definite diffusion tensor, and α, β, γ are parameters of boundary conditions on domain boundary $\partial\Omega$ with outer normal unit vector $\mathbf{n}$. Combination $\alpha = 1, \beta = 0$ represents Dirichlet boundary conditions, and combination $\alpha = 0, \beta = 1$ represents Neumann boundary conditions. If parameters α, β are functions of coordinates $\mathbf{x}$, then the boundary condition is mixed: on $\Gamma_D = \overline{\Gamma_D}$ Dirichlet condition is imposed and the trace of p on Γ_D is equal to $\gamma(\mathbf{x})$. On $\Gamma_N = \partial\Omega \setminus \Gamma_D$, Neumann boundary condition is imposed, $\mathbf{n} \cdot (\mathbb{K}\nabla p) = \gamma(\mathbf{x})$. The set of boundary mesh faces $\mathcal{F}_B$ is split into subsets $\mathcal{F}_D$ and $\mathcal{F}_N$ according to their location at Γ_D or Γ_N.

The cell-centered finite volume (FV) method is based on Stokes theorem applied to the divergence operator in (2.1) in each cell ω:

$$-\int_{\omega} \mathrm{div}\,(\mathbb{K}\nabla p)\,\mathrm{d}\omega = -\oint_{\partial\omega} \mathbf{n} \cdot (\mathbb{K}\nabla p)\,\mathrm{d}s = \sum_{f \in \mathcal{F}(\omega)} -\int_{f} \mathbf{n} \cdot (\mathbb{K}\nabla p)\,\mathrm{d}s = \sum_{f \in \mathcal{F}(\omega)} q|f|, \tag{2.2}$$

where $\partial\omega$ is the boundary of cell ω and $\mathcal{F}(\omega)$ is the set of faces of ω, $|f|$ is the area of face f, $q|f| = -\int_f \mathbf{n} \cdot (\mathbb{K}\nabla p)\,\mathrm{d}s$ is the total flux across f. The cornerstone of the finite volume method is approximation of the averaged flux density q on internal and boundary faces. For the sake of brevity, we shall refer to q as flux if it does not bring ambiguity. The approximation q_h of q determines the properties of the finite volume method.

The first step to flux discretization is conventional. Using the symmetry of $\mathbb{K}$, we can flip the position of the tensor within the scalar product as follows: $-\mathbf{n} \cdot (\mathbb{K}\nabla p) = -\mathbb{K}\mathbf{n} \cdot \nabla p$. This suggests approximation of flux $q|f|$ based on approximation of the gradient ∇p along the co-normal direction $\boldsymbol{\ell} = \mathbb{K}\mathbf{n}$.

The second step is derivation of flux discretization formula. We assume that the discrete solution is collocated at the centers of mesh cells and the direction of $\mathbf{n}_f$ for face f is fixed. Denote by $\mathcal{P}_f$ the set of cells contributing values p_j of the discrete solution to the formula of the discrete flux computation for face f

$$q_h = \sum_{\omega_j \in \mathcal{P}_f} T_{f,j}\, p_j + G_f, \tag{2.3}$$

where G_f denotes possible contributions of the boundary conditions.

The cell-centered FV method discretizes equations (2.1) in each cell ω_i, $i = 1, \ldots, N$, $N = \#C(\Omega_h)$ replacing the flux q with its numerical approximation (2.3):

$$\sum_{f \in \mathcal{F}(\omega_i)} |f|\chi_{\omega_i,f}\, q_h = \sum_{f \in \mathcal{F}(\omega_i)} |f|\chi_{\omega_i,f} \left(\sum_{\omega_j \in \mathcal{P}_f} T_{f,j}\, p_j + G_f \right) = \int_{\omega_i} g\,\mathrm{d}\omega, \tag{2.4}$$

where $\chi_{\omega_i,f}$ is either 1 or -1 depending on the mutual orientation of outer normal for ω_i and $\mathbf{n}_f$.

The equalities (2.4) define the system of N algebraic equations with N unknowns p_j, $j = 1, \ldots, N$. If coefficients $T_{f,j}$ do not depend on p_j, then the above equations are linear and (2.4) may be written in the matrix form:

$$\mathbb{A}\mathbf{p} = \mathbf{g}, \tag{2.5}$$

where vector $\mathbf{p}$ is composed of p_j, and matrix $\mathbb{A}$ is assembled from coefficients in (2.3) multiplied by $|f|\chi_{\omega_i,f}$. Matrix $\mathbb{A}$ is sparse, and each row i contains non-zero entries with column indices j corresponding to cells ω_j from $\mathcal{P}_f$. The system can be solved by direct sparse factorization methods [21, 22, 52] or Krylov subspace iterative methods [121, 128].

The sparsity pattern for $\mathbb{A}$ depends on the set $\mathcal{P}_f$. For instance, if the discrete flux q_h has the two-point stencil with coefficients T_+, T_-, i.e., $\mathcal{P}_f = \{\omega_+; \omega_-\}$ for an internal face f shared by two cells ω_+ and ω_-, then the assembling implies $\mathbb{A}_{ij} := \mathbb{A}_{ij} + [\mathbb{D}_f\mathbb{T}_f]_\pm$ with 2×2 matrices

$$\mathbb{T}_f = \begin{pmatrix} T_+ & -T_- \\ -T_+ & T_- \end{pmatrix}, \quad \mathbb{D}_f = \begin{pmatrix} |f|\chi_{\omega_+,f} & 0 \\ 0 & |f|\chi_{\omega_-,f} \end{pmatrix}, \tag{2.6}$$

where i and j are indexes of cells ω_+ and ω_- in a global numeration of mesh cells, and $[\mathbb{D}_f\mathbb{T}_f]_\pm$ are entries of $\mathbb{D}_f\mathbb{T}_f$ corresponding to cells ω_+ and ω_-. The sparsity pattern for each row of matrix $\mathbb{A}$ is formed by the closest (i.e., sharing a face) neighboring cells.

If coefficients $T_{f,j}$ depend on $\mathbf{p}$, then (2.4) is the system of nonlinear equations:

$$\mathbb{A}(\mathbf{p})\mathbf{p} = \mathbf{g}(\mathbf{p}), \tag{2.7}$$

where entries of matrix $\mathbb{A}$ and right-hand side $\mathbf{g}$ may depend on $\mathbf{p}$.

The solution of (2.7) (if exists) can be found by available nonlinear solvers:

- Picard iteration $\mathbb{A}(\mathbf{p}^k)\mathbf{p}^{k+1} = \mathbf{g}(\mathbf{p}^k)$;
- Anderson acceleration [109] of Picard iterations;
- Newton method with analytically computed Jacobian or Jacobian due to automated differentiation.

Although nonlinear flux discretizations produce nonlinear systems even for linear PDEs, they provide both approximation and monotonicity of the discrete solution. Depending on the scheme, by monotonicity we understand either discrete solution positivity or the discrete maximum principle. Numerical and theoretical analysis of the schemes is given at the end of the chapter.

Now, we proceed to derivation of flux discretizations (2.3). The derivation is based on definition of the discrete solution and assumptions for problem and mesh data. We present two approaches to derivation: finite difference approximation for collocated discrete solution and gradient recovery of piecewise linear local approximation of the solution.

We distinguish discretization for internal and boundary faces. An internal face f is assumed to be shared by two cells ω_+ and ω_- with barycenters $\mathbf{x}_+$, $\mathbf{x}_-$ and diffusion tensors $\mathbb{K}_+$, $\mathbb{K}_-$, respectively. A boundary face f is assumed to belong to cell ω_+ with diffusion tensor $\mathbb{K}_+$.

2.2 Monotone Two-Point Flux Approximation Based on Finite Differences

First, one assumes that discrete solutions p_* are collocated at barycenters $\mathbf{x}_*$ of cells ω_*. Therefore, approximate derivative of the discrete solution may be computed by finite differences. If one assumes a homogeneous media and mesh $\mathbb{K}$-orthogonality (i.e., collinearity of $\boldsymbol{\ell}$ and $\mathbf{x}_+ - \mathbf{x}_-$), then one defines the linear two-point flux approximation (TPFA)

$$q_h = \frac{\boldsymbol{\ell} \cdot (\mathbf{x}_+ - \mathbf{x}_-)}{|\mathbf{x}_+ - \mathbf{x}_-|} \frac{p_+ - p_-}{|\mathbf{x}_+ - \mathbf{x}_-|}, \tag{2.8}$$

where $q = q_h + O(h)$, $h = |\mathbf{x}_+ - \mathbf{x}_-|$.

If the media is homogeneous and the mesh is not $\mathbb{K}$-orthogonal, then (2.8) does not provide approximation, i.e., $q_h \not\to q$ as $h \to 0$. In this case, we will design a nonlinear flux approximation. To this end, for every cell ω we define a set Σ_ω of nearby collocation points as follows. First, we add to Σ_ω the collocation point $\mathbf{x}_\omega$. Then, for every face $f \in \mathcal{F}(\omega)$, we add the collocation point $\mathbf{x}_{\omega'_f}$, where ω'_f is the cell, other than ω, that has face f.

Returning to internal face f shared by two cells ω_+ and ω_- with the unit normal $\mathbf{n}$ directed toward ω_-, we assume that there exist three points $\mathbf{x}_{\pm,1}$, $\mathbf{x}_{\pm,2}$, and $\mathbf{x}_{\pm,3}$ in set $\Sigma_{\omega_\pm}$, that for vectors $\mathbf{t}_{\pm,k} = \mathbf{x}_{\pm,k} - \mathbf{x}_\pm$, $k = 1, 2, 3$, the following condition holds:

$$\boldsymbol{\ell} = \alpha_{+,1}\mathbf{t}_{+,1} + \alpha_{+,2}\mathbf{t}_{+,2} + \alpha_{+,3}\mathbf{t}_{+,3}, \tag{2.9}$$

$$-\boldsymbol{\ell} = \alpha_{-,1}\mathbf{t}_{-,1} + \alpha_{-,2}\mathbf{t}_{-,2} + \alpha_{-,3}\mathbf{t}_{-,3}, \tag{2.10}$$

where $\alpha_{\pm,1} > 0$, $\alpha_{\pm,2} \geq 0$, and $\alpha_{\pm,3} \geq 0$. In geometric terms, this implies that the co-normal vector $\boldsymbol{\ell}$ started from $\mathbf{x}_+$ belongs to the trihedral corner formed by vectors $\mathbf{t}_{+,k}$, whereas the co-normal vector $-\boldsymbol{\ell}$ started from $\mathbf{x}_-$ belongs to the trihedral corner formed by vectors $\mathbf{t}_{-,k}$, see Fig. 2.1 for 2D example.

We also assume that if one of the points $\mathbf{x}_{\pm,k}$ coincides with $\mathbf{x}_\mp$, its index $k = 1$. Note that if $\Sigma_{\omega_\pm}$ does not contain the desired three points, one can extend the set $\Sigma_{\omega_\pm}$ with other neighbors of $\omega_\pm$ increasing the discretization stencil.

Now, using

$$\boldsymbol{\ell} \cdot \nabla p = \sum_{k=1,2,3} \alpha_{+,k} \nabla p \cdot \mathbf{t}_{+,k}, \quad -\boldsymbol{\ell} \cdot \nabla p = \sum_{k=1,2,3} \alpha_{-,k} \nabla p \cdot \mathbf{t}_{-,k}$$

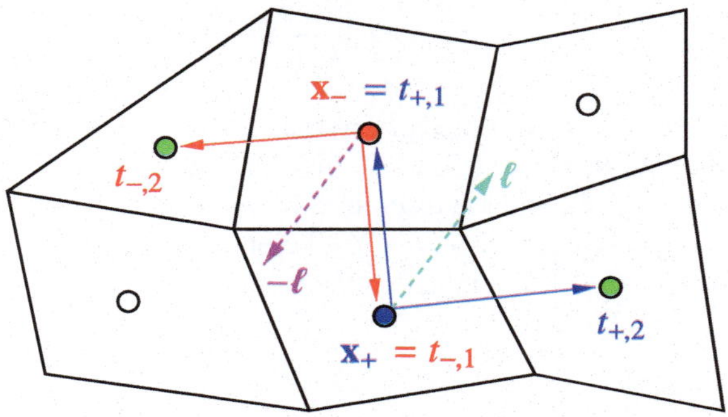

Fig. 2.1 Two representations of co-normal vector $\boldsymbol{\ell}$ that belongs to the angle formed by $\mathbf{t}_{+,k}$ and $\mathbf{t}_{-,k}$, respectively (2D example)

and finite differences to approximate directional derivatives, we obtain two approximations of flux q:

$$
\begin{aligned}
q_h^+ &= -\sum_{k=1,2,3} \alpha_{+,k}(p_{+,k} - p_+), \\
q_h^- &= \sum_{k=1,2,3} \alpha_{-,k}(p_{-,k} - p_-).
\end{aligned} \tag{2.11}
$$

Note that $q = q_h^+ + O(h), q = q_h^- + O(h)$. The final numerical flux is a linear combination of these two fluxes:

$$
\begin{aligned}
q_h &= \mu_+ q_h^+ + \mu_-(-q_h^-) \\
&= \mu_+(\alpha_{+,1}(p_+ - p_{+,1}) + \alpha_{+,2}(p_+ - p_{+,2}) + \alpha_{+,3}(p_+ - p_{+,3})) \\
&- \mu_-(\alpha_{-,1}(p_- - p_{-,1}) + \alpha_{-,2}(p_- - p_{-,2}) + \alpha_{-,3}(p_- - p_{-,3})).
\end{aligned} \tag{2.12}
$$

The weights μ_+ and μ_- are selected to keep only two unknowns p_+ and p_- in (2.12). The second requirement is to approximate the true flux. These requirements lead us to the following system:

$$
\begin{aligned}
\mu_+ d_+ - \mu_- d_- &= 0, \\
\mu_+ + \mu_- &= 1,
\end{aligned} \tag{2.13}
$$

where $d_\pm = (\alpha_{\pm,1} p_{\pm,1} + \alpha_{\pm,2} p_{\pm,2} + \alpha_{\pm,3} p_{\pm,3})$, if $\mathbf{x}_\mp \neq \mathbf{x}_{\pm,1}$, and $d_\pm = (\alpha_{\pm,2} p_{\pm,2} + \alpha_{\pm,3} p_{\pm,3})$ otherwise. We also define $\beta_\pm = \alpha_{\pm,1}$ if $\mathbf{x}_\mp = \mathbf{x}_{\pm,1}$, and $\beta_\pm = 0$, otherwise. The solution of (2.13), $\mu_+ = d_-/(d_+ + d_-)$, $\mu_- = d_+/(d_+ + d_-)$ defines the nonlinear two-point flux approximation

$$
q_h = T_+ p_+ - T_- p_-, \quad T_\pm = \mu_\pm |\boldsymbol{\ell}| \sum_{k=1}^{3} \alpha_{\pm,k} |\mathbf{t}_{+,k}| + \mu_\mp \, |\boldsymbol{\ell}| \, \beta_\pm \, |\mathbf{t}_{+,k}|. \tag{2.14}
$$

To avoid possible division by 0 in case $d_+ + d_- = 0$, we modify $\mu_\pm$ by regularization

$$\mu_+ = \frac{d_- + \varepsilon}{d_+ + d_- + 2\varepsilon}, \quad \mu_- = \frac{d_+ + \varepsilon}{d_+ + d_- + 2\varepsilon}, \tag{2.15}$$

where small $\varepsilon > 0$ is equal to the machine precision.

Note that coefficients $T_\pm$ depend on problem coefficients ($\mathbb{K}$), mesh geometry ($\alpha_\pm, \mathbf{t}_{\pm,k}, \boldsymbol{\ell}$), and the discrete solution ($\mu_\pm$).

If a boundary face $f \in \mathcal{F}_N$ belongs to the Neumann part of the boundary where $\alpha = 0$, $\beta = 1$, then the flux is given as

$$q_h = \gamma. \tag{2.16}$$

If a boundary face $f \in \mathcal{F}_D \cap \mathcal{F}(\omega_+)$, i.e., belongs to the Dirichlet part of the boundary where $\alpha = 1$, $\beta = 0$, then we exploit face f as a slim cell ω_- with collocated value $p_f = \frac{1}{|f|}\int\limits_f \gamma ds$ at its center $\mathbf{x}_f$ and apply the same formula (2.14). Note that vectors from the triplet $\{\mathbf{t}_{-,1}, \mathbf{t}_{-,2}, \mathbf{t}_{-,3}\}$ may be chosen from vectors $\mathbf{x}_+ - \mathbf{x}_f$, $\mathbf{x}_{e_k} - \mathbf{x}_f$, $k = 1, 2, 3$, where $\mathbf{x}_{e_k}$ is the center of edge e_k of face f, and $p(\mathbf{x}_{e_k}) = \frac{1}{|e_k|}\int\limits_{e_k} \gamma de$.

For inhomogeneous media, the derivation of flux discretization is based on interpolated values in cell vertices and/or in face centers [51]. We shall not present these approaches since we are going to present the most general approach for derivation of nonlinear TPFA.

2.3 Monotone Two-Point Flux Approximation Based on Gradient Reconstruction

In the general case of piecewise-constant coefficient $\mathbb{K}$, we assume additionally that jumps of $\mathbb{K}$ may occur across faces of Ω_h. Moreover, in a vicinity of such face f the discrete solution p_h may be recovered from the collocated values to a continuous piecewise linear function which is represented by linear functions $p_\pm(\mathbf{x})$ in cells $\omega_\pm$ sharing f. This implies that the gradient ∇p_h is constant in each cell $\omega_\pm$. To calculate the fluxes $q_+ = -\mathbb{K}_+\mathbf{n} \cdot \nabla p_+$ and $q_- = -\mathbb{K}_-\mathbf{n} \cdot \nabla p_-$, we need gradient reconstruction which uses boundary data and an interpolated value at a point $\mathbf{x}_*$ belonging to the plane of face f.

The divergence form of (2.1) suggests discrete flux continuity on face f

$$-\mathbb{K}_+\mathbf{n} \cdot \nabla p_+ = -\mathbb{K}_-\mathbf{n} \cdot \nabla p_-, \tag{2.17}$$

whereas continuity of p_h at point $\mathbf{x}_*$ suggests

$$p_+ + (\mathbf{x}_* - \mathbf{x}_+) \cdot \nabla p_+ = p_- + (\mathbf{x}_* - \mathbf{x}_-) \cdot \nabla p_-. \tag{2.18}$$

Let us define the distances $r_+ = \mathbf{n} \cdot (\mathbf{x}_* - \mathbf{x}_+)$ and $r_- = \mathbf{n} \cdot (\mathbf{x}_- - \mathbf{x}_*)$. If the cell $\omega_\pm$ is star-shaped with respect to $\mathbf{x}_\pm$, then $r_\pm > 0$. For general non-convex cell, one may need to change the collocation point to satisfy $r_\pm > 0$. Using this notation and $p_* = p_h(\mathbf{x}_*)$, we express the gradients by

$$\begin{aligned} \nabla p_+ &= \frac{1}{r_+}\mathbf{n}\,(p_* - p_+) + \left(\mathbb{I} - \frac{1}{r_+}\mathbf{n} \cdot (\mathbf{x}_* - \mathbf{x}_+)\right) \nabla p_\tau, \\ \nabla p_- &= \frac{1}{r_-}\mathbf{n}\,(p_- - p_*) + \left(\mathbb{I} - \frac{1}{r_-}\mathbf{n} \cdot (\mathbf{x}_- - \mathbf{x}_*)\right) \nabla p_\tau, \end{aligned} \tag{2.19}$$

where $\mathbb{I}$ is the unit matrix of order 3, $\nabla p_\tau = \left(\mathbb{I} - \mathbf{n}\mathbf{n}^T\right) \nabla p_+ = \left(\mathbb{I} - \mathbf{n}\mathbf{n}^T\right) \nabla p_-$ is the transversal part of the gradients. The latter has to be continuous due to continuity of p_h. Using (2.19) and (2.17), we get

$$\begin{aligned} p_* = &\frac{\lambda_+ r_- p_+ + \lambda_- r_+ p_-}{\lambda_+ r_- + \lambda_- r_+} \\ &+ \left(\mathbf{x}_* - \frac{r_+ r_- \left(\mathbb{K}_+ - \mathbb{K}_-\right)\mathbf{n} + \lambda_+ r_- \mathbf{x}_+ + \lambda_- r_+ \mathbf{x}_-}{\lambda_+ r_- + \lambda_- r_+}\right) \cdot \nabla p_\tau, \end{aligned} \tag{2.20}$$

where the co-normal projection $\lambda_\pm = \mathbf{n}^T \mathbb{K}_\pm \mathbf{n}$ is positive due to positive-definiteness of $\mathbb{K}_\pm$. We see that there exists such point

$$\mathbf{x}_H = \frac{r_+ r_- \left(\mathbb{K}_+ - \mathbb{K}_-\right)\mathbf{n} + \lambda_+ r_- \mathbf{x}_+ + \lambda_- r_+ \mathbf{x}_-}{\lambda_+ r_- + \lambda_- r_+} \tag{2.21}$$

that for $\mathbf{x}_* = \mathbf{x}_H$ the expression for p_* has the two-point stencil:

$$p_H = \frac{\lambda_+ r_- p_+ + \lambda_- r_+ p_-}{\lambda_+ r_- + \lambda_- r_+}. \tag{2.22}$$

Such interpolation point $\mathbf{x}_H$ is known as the harmonic averaging point, and interpolation (2.22) is known as harmonic averaging [19].

Let now f be a boundary face adjacent to cell ω_+ (i.e., $f = \omega_+ \cap \partial\Omega$) with outer unit normal $\mathbf{n}$ and face center $\mathbf{x}_b$. Let also $\mathbf{x}_*$ be a point belonging to the plane of face f. Using the expression for ∇p_+ from (2.19) and $p(\mathbf{x}_b) = p_* + (\mathbf{x}_b - \mathbf{x}_*)^T \nabla p_\tau$ in boundary conditions from (2.1), we obtain an expression for p_* at $\mathbf{x}_*$:

$$\begin{aligned} p_* = &\left(\alpha + \beta\frac{\lambda_+}{r_+}\right)^{-1} \left(\gamma + \beta\frac{\lambda_+}{r_+} p_+\right) \\ &+ \left(\mathbf{x}_* - \left(\alpha + \beta\frac{\lambda_+}{r_+}\right)^{-1} \left(\beta\mathbb{K}_+\mathbf{n} + \beta\frac{\lambda_+}{r_+}\mathbf{x}_+ + \alpha\mathbf{x}_b\right)\right) \cdot \nabla p_\tau. \end{aligned} \tag{2.23}$$

Similar to (2.20), we see that there exists such point

$$\mathbf{x}_H = \left(\alpha + \beta \frac{\lambda_+}{r_+}\right)^{-1} \left(\beta \mathbb{K}_+ \mathbf{n} + \beta \frac{\lambda_+}{r_+} \mathbf{x}_+ + \alpha \mathbf{x}_b\right)$$

that ∇p_τ is not involved in interpolation of p_*:

$$p_H = \left(\alpha + \beta \frac{\lambda_+}{r_+}\right)^{-1} \left(\gamma + \beta \frac{\lambda_+}{r_+} p_+\right). \tag{2.24}$$

Now we can proceed to gradient ∇p_+ reconstruction in cell ω_+. Reconstruction of ∇p_- in cell ω_- is derived similarly. For any face $f \in \mathcal{F}(\omega_+)$ of the cell ω_+, one has

$$(\mathbf{x}_H - \mathbf{x}_+) \cdot \nabla p_+ = p_H - p_+. \tag{2.25}$$

For internal face $f = \omega_+ \cap \omega_-$, we use (2.22):

$$\begin{aligned} p_H &= (1 - m_H) p_+ + m_H p_-, \quad \mathbf{x}_H = (1 - m_H)\mathbf{x}_+ + m_H \mathbf{x}_- + \boldsymbol{\ell}_H, \\ m_H &= \frac{\lambda_- r_+}{\lambda_+ r_- + \lambda_- r_+}, \quad \boldsymbol{\ell}_H = \frac{r_+ r_- \left(\mathbb{K}_+ - \mathbb{K}_-\right) \mathbf{n}}{\lambda_+ r_- + \lambda_- r_+}, \end{aligned} \tag{2.26}$$

and for boundary face $f = \omega_+ \cap \partial\Omega$ we use (2.24):

$$\begin{aligned} p_H &= (1 - m_H) p_+ + g_H, \quad \mathbf{x}_H = (1 - m_H)\mathbf{x}_+ + \boldsymbol{\ell}_H, \\ m_H &= \left(\alpha + \beta \frac{\lambda_+}{r_+}\right)^{-1} \alpha, \quad \boldsymbol{\ell}_H = \left(\alpha + \beta \frac{\lambda_+}{r_+}\right)^{-1} \left(\beta \mathbb{K}_+ \mathbf{n} + \alpha \mathbf{x}_b\right), \\ g_H &= \left(\alpha + \beta \frac{\lambda_+}{r_+}\right)^{-1} \gamma. \end{aligned} \tag{2.27}$$

If a non-convex cell ω_+ is star-shaped with respect to its center, coefficients m_H in (2.26) and (2.27) belong to interval $(0, 1)$.

In order to reconstruct ∇p_+, we choose three faces $f_1, f_2, f_3 \in \mathcal{F}(\omega_+)$ and solve the system:

$$\begin{pmatrix} (\mathbf{x}_{H_1} - \mathbf{x}_+)^T \\ (\mathbf{x}_{H_2} - \mathbf{x}_+)^T \\ (\mathbf{x}_{H_3} - \mathbf{x}_+)^T \end{pmatrix} \nabla p_+ = \begin{pmatrix} p_{H_1} - p_+ \\ p_{H_2} - p_+ \\ p_{H_3} - p_+ \end{pmatrix}. \tag{2.28}$$

The inverse Q of the matrix in (2.28) is obtained by

$$\begin{aligned} Q &= \begin{pmatrix} (\mathbf{x}_{H_1} - \mathbf{x}_+)^T \\ (\mathbf{x}_{H_2} - \mathbf{x}_+)^T \\ (\mathbf{x}_{H_3} - \mathbf{x}_+)^T \end{pmatrix}^{-1} \\ &= \frac{\left((\mathbf{x}_{H_2} - \mathbf{x}_+) \times (\mathbf{x}_{H_3} - \mathbf{x}_+) \ (\mathbf{x}_{H_1} - \mathbf{x}_+) \times (\mathbf{x}_{H_3} - \mathbf{x}_+) \ (\mathbf{x}_{H_1} - \mathbf{x}_+) \times (\mathbf{x}_{H_2} - \mathbf{x}_+) \right)}{(\mathbf{x}_{H_1} - \mathbf{x}_+) \cdot \left((\mathbf{x}_{H_2} - \mathbf{x}_+) \times (\mathbf{x}_{H_3} - \mathbf{x}_+) \right)}. \end{aligned} \tag{2.29}$$

For a boundary face $f = \omega_+ \cap \partial\Omega$, one needs to reconstruct also the gradient at face center $\mathbf{x}_b$ where unknown p_b is collocated. Then, instead of (2.28), we write

$$(\mathbf{x}_+ - \mathbf{x}_b) \cdot \nabla p_b = p_+ - p_b, \tag{2.30}$$

and additional conditions on two faces $f_i \in \mathcal{F}(\omega_+)$, $f_i \neq f$:

$$\left(\mathbf{x}_{H_i} - \mathbf{x}_b\right) \cdot \nabla p_b = p_{H_i} - p_b. \tag{2.31}$$

In this case in expressions (2.21), (2.22) for $\mathbf{x}_{H_i}$ and p_{H_i}, we use $\mathbf{x}_b$ instead of $\mathbf{x}_+$, p_b instead of p_+ and distance from face f_i to $\mathbf{x}_b$ instead of distance r_+. As we will further see, the actual value of p_b will not be used.

Computation of the co-normal $\boldsymbol{\ell}$ for face f of cell ω_+ exploits matrix Q from (2.29):

$$\mathbb{K}_+\mathbf{n} \cdot \nabla p_+ = \mathbb{K}_+\mathbf{n} \cdot Q \begin{pmatrix} p_{H_1} - p_+ \\ p_{H_2} - p_+ \\ p_{H_3} - p_+ \end{pmatrix} = \left[c_1 \; c_2 \; c_3 \right] \begin{pmatrix} p_{H_1} - p_+ \\ p_{H_2} - p_+ \\ p_{H_3} - p_+ \end{pmatrix}, \tag{2.32}$$

where c_1, c_2, c_3 are the coefficients of the stencil. For both internal and boundary fluxes, we search for such faces $f_1, f_2, f_3 \in \mathcal{F}(\omega_+)$ that c_1, c_2, c_3 are non-negative and their sum is minimal. If there exist several admissible triplets of f_1, f_2, f_3, we select the one for which condition number of matrix $Q^T \mathbb{K}_+ Q$ is minimal. The stencil has to be computed only once for a given grid and coefficients; thus, the computational cost of stencil selection is not of a big concern.

Now we can proceed to the flux discretization on internal face $f = \omega_+ \cap \omega_-$. To present the general case, we assume that the three faces $f_{+,1}, f_{+,2}, f_{+,3} \in \mathcal{F}(\omega_+)$ selected for calculation of ∇p_+ in cell ω_+ with coefficients $c_{+,1}, c_{+,2}, c_{+,3}$ correspond to the face f, another internal face, and a boundary face, respectively. Then $f_{+,1} = \omega_+ \cap \omega_-$, $f_{+,2} = \omega_+ \cap \omega_{+,2}$, $f_3 = \omega_+ \cap \partial\Omega$. Analogously, we assume that the three faces $f_{-,1}, f_{-,2}, f_{-,3} \in \mathcal{F}(\omega_-)$ selected for calculation of ∇p_- in cell ω_- with coefficients $c_{-,1}, c_{-,2}, c_{-,3}$ correspond to the face f, another internal face, and a boundary face, respectively, and $f_{-,1} = \omega_- \cap \omega_+$, $f_{-,2} = \omega_- \cap \omega_{-,2}$, $f_{-,3} = \omega_- \cap \partial\Omega$. Then, the discrete fluxes q_h^+ and q_h^- are expressed as

$$\begin{aligned} q_h^+ &= c_{+,1} m_{+,1} \left(p_+ - p_-\right) + c_{+,2} m_{+,2} \left(p_+ - p_{+,2}\right) + c_{+,3} \left(m_{+,3} p_+ - g_{+,3}\right), \\ q_h^- &= c_{-,1} m_{-,1} \left(p_+ - p_-\right) + c_{-,2} m_{-,2} \left(p_{-,2} - p_-\right) + c_{-,3} \left(g_{-,3} - m_{-,3} p_-\right), \end{aligned} \tag{2.33}$$

where $m_{\pm,1}, m_{\pm,2}$ are defined by m_H in (2.26) and $m_{\pm,3}, g_{\pm,3}$ are defined by m_H and g_H in (2.27).

We define the convex ($\mu_+ + \mu_- = 1$) combination of fluxes with positive coefficients μ_+ and μ_-:

$$\begin{aligned} q_h = \mu_+ q_h^+ + \mu_- q_h^- &= \left(\mu_+ \left(c_{+,1}m_{+,1} + c_{+,2}m_{+,2} + c_{+,3}m_{+,3}\right) + \mu_- c_{-,1}m_{-,1}\right) p_+ \\ &- \left(\mu_- \left(c_{-,1}m_{-,1} + c_{-,2}m_{-,2} + c_{-,3}m_{-,3}\right) + \mu_+ c_{+,1}m_{+,1}\right) p_- \\ &+ \mu_- \left(c_{-,2}m_{-,2}p_{-,2} + c_{-,3}g_{-,3}\right) - \mu_+ \left(c_{+,2}m_{+,2}p_{+,2} + c_{+,3}g_{+,3}\right) \\ &= T_+ p_+ - T_- p_- + \mu_- R_- - \mu_+ R_+ + G, \end{aligned} \tag{2.34}$$

with the definitions:

$$\begin{aligned} T_+ &= \mu_+ \left(c_{+,1}m_{+,1} + c_{+,2}m_{+,2} + c_{+,3}m_{+,3}\right) + \mu_- c_{-,1}m_{-,1} \geq 0, \\ T_- &= \mu_- \left(c_{-,1}m_{-,1} + c_{-,2}m_{-,2} + c_{-,3}m_{-,3}\right) + \mu_+ c_{+,1}m_{+,1} \geq 0, \\ R_+ &= c_{+,2}m_{+,2}p_{+,2} \geq 0, \\ R_- &= c_{-,2}m_{-,2}p_{-,2} \geq 0, \\ G &= \mu_- c_{-,3}g_{-,3} - \mu_+ c_{+,3}g_{+,3}, \end{aligned} \tag{2.35}$$

coefficients T_+, T_-, R_+, R_-, and G depend on unknowns $p_\pm$, $p_{\pm,2}$ in the stencil.

The solution $\mu_+ = R_-/(R_+ + R_-)$, $\mu_- = R_+/(R_+ + R_-)$ to

$$\begin{aligned} \mu_+ R_+ - \mu_- R_- &= 0, \\ \mu_+ + \mu_- &= 1 \end{aligned} \tag{2.36}$$

produces the two-point flux approximation q_h:

$$q_h = T_+ p_+ - T_- p_- + G. \tag{2.37}$$

To avoid degeneracy in case $R_+ + R_- = 0$, we use regularized solution

$$\mu_+ = \frac{R_- + \varepsilon}{R_+ + R_- + 2\varepsilon}, \quad \mu_- = \frac{R_+ + \varepsilon}{R_+ + R_- + 2\varepsilon}, \tag{2.38}$$

where small $\varepsilon > 0$ is the machine precision. Note that T_+ and T_- are continuously differentiable with respect to unknowns from the stencil provided that the unknowns are positive.

Now we discuss the flux discretization on boundary face $f = \omega_+ \cap \partial\Omega$. For the general case, we assume that the three faces $f_{+,1}, f_{+,2}, f_{+,3} \in \mathcal{F}(\omega_+)$ selected for calculation of ∇p_+ in cell ω_+ with coefficients $c_{+,1}, c_{+,2}, c_{+,3}$ correspond to face f, an internal face, and another boundary face of cell ω_+, respectively. Then $f_{+,1} = f$, $f_{+,2} = \omega_+ \cap \omega_{+,2}$, $f_3 = \omega_+ \cap \partial\Omega \neq f$. Analogously, we assume that the three faces $f_{-,1}, f_{-,2}, f_{-,3} \in \mathcal{F}(\omega_+)$ selected for calculation of gradient at face center $\mathbf{x}_b$ (where unknown p_b is collocated) with coefficients $c_{-,1}, c_{-,2}, c_{-,3}$ correspond to face f, an internal face, and another boundary face of cell ω_+, respectively, and $f_{-,1} = f$, $f_{-,2} = \omega_+ \cap \omega_{+,2}$, $f_{-,3} = \omega_+ \cap \partial\Omega \neq f$. Then the discrete fluxes q_h^+ and q_h^- are expressed as

$$\begin{aligned} q_h^+ &= c_{+,1}m_{+,1}\left(p_+ - p_b\right) + c_{+,2}m_{+,2}\left(p_+ - p_{+,2}\right) + c_{+,3}\left(m_{+,3}p_+ - g_{+,3}\right), \\ q_h^- &= c_{-,1}m_{-,1}\left(p_+ - p_b\right) + c_{-,2}m_{-,2}\left(p_{-,2} - p_b\right) + c_{-,3}\left(g_{-,3} - m_{-,3}p_b\right), \end{aligned} \tag{2.39}$$

where $m_{\pm,2}$ are defined by m_H in (2.26) and $m_{\pm,1}, m_{\pm,3}, g_{\pm,3}$ are defined by m_H and g_H in (2.27).

Using the convex combination, we get

$$\begin{aligned} q_h = \mu_+ q_h^+ + \mu_- q_h^- &= \left(\mu_+ \left(c_{+,1}m_{+,1} + c_{+,2}m_{+,2} + c_{+,3}m_{+,3}\right) + \mu_- c_{-,1}m_{-,1}\right) p_+ \\ &- \left(\mu_- \left(c_{-,1}m_{-,1} + c_{-,2}m_{-,2} + c_{-,3}m_{-,3}\right) + \mu_+ c_{+,1}m_{+,1}\right) p_b \\ &+ \mu_- \left(c_{-,2}m_{-,2}p_{-,2} + c_{-,3}g_{-,3}\right) - \mu_+ \left(c_{+,2}m_{+,2}p_{+,2} + c_{+,3}g_{+,3}\right) \\ &= T_+ p_+ - T_- p_b + \mu_- R_- - \mu_+ R_+ + G, \end{aligned} \tag{2.40}$$

with the definitions for T_+, T_-, R_+, R_-, G similar to (2.35) and defining μ_+ and μ_- similarly to (2.38) we finally arrive to

$$q_h = T_+ p_+ - T_- p_b + G. \tag{2.41}$$

We can eliminate p_b from (2.41) by considering the boundary conditions:

$$\alpha p_b + \beta \left(T_- p_b - T_+ p_+ - G\right) = \gamma, \quad p_b = \left(\alpha + \beta T_-\right)^{-1} \left(\beta T_+ p_+ + \gamma + \beta G\right), \tag{2.42}$$

then with (2.42) in (2.41) we get

$$q_h = \left(\alpha + \beta T_-\right)^{-1} \left(\alpha \left(T_+ p_+ + G\right) - T_- \gamma\right). \tag{2.43}$$

Expression (2.43) does not require p_b. Expressions (2.37) and (2.43) are expressions for the flux discretization.

2.4 Monotone Multi-Point Flux Approximation

The two-point flux approximations (2.14) and (2.37) admit multi-point generalizations which result in schemes satisfying the discrete maximum principle excluding spurious oscillations in the numerical solution.

The alternative to (2.13) or (2.36) requirement to the weights μ_+ and μ_- balances the relative contribution of the left and the right fluxes to the final flux. The second requirement remains approximation of the true flux. These requirements lead us to the following system:

$$\begin{aligned} q_h^+ \mu_+ + q_h^- \mu_- &= 0, \\ \mu_+ + \mu_- &= 1. \end{aligned} \tag{2.44}$$

We must consider two cases. In the first case, $q_h^+ q_h^- \le 0$ and the regularized solution to (2.44) is

$$\mu_+ = \frac{|q_h^-| + \varepsilon}{|q_h^+| + |q_h^-| + 2\varepsilon}, \quad \mu_- = \frac{|q_h^+| + \varepsilon}{|q_h^+| + |q_h^-| + 2\varepsilon}. \tag{2.45}$$

Thus, the flux discretization has two equivalent algebraic representations:

$$q_h = 2q_h^+ = -2q_h^-. \tag{2.46}$$

In case of one-sided flux approximations (2.11), the discrete flux (2.46) becomes

$$\begin{aligned} q_h &= 2\mu_\pm(\alpha_{\pm,1}(p_\pm - p_{\pm,1}) + \alpha_{\pm,2}(p_\pm - p_{\pm,2}) + \alpha_{\pm,3}(p_\pm - p_{\pm,3})) \\ &= T_{\pm,1}(p_\pm - p_{\pm,1}) + T_{\pm,2}(p_\pm - p_{\pm,2}) + T_{\pm,3}(p_\pm - p_{\pm,3}) \end{aligned} \tag{2.47}$$

with non-negative coefficients $T_{\pm,i} = 2\mu_\pm\alpha_{\pm,i}$. Each representation has the zero row sum and contributes to matrix $\mathbb{A}$ assembling for the cell associated with it. Note that these coefficients depend on the fluxes and hence on the discrete solution in neighboring cells.

The second case $q_h^+ q_h^- > 0$ leads to a potentially degenerate diffusive flux. In order to avoid this degeneracy, we re-group the terms in (2.12)

$$q_h = \mu_+\tilde{q}_h^+ + \mu_-(-\tilde{q}_h^-) + (\mu_+\alpha_{+,1} + \mu_-\alpha_{-,1})(p_+ - p_-), \tag{2.48}$$

where $\tilde{q}_h^\pm$ collects all the remaining terms. In case of discretization (2.11) $\tilde{q}_h^+ = \alpha_{+,2}(p_+ - p_{+,2}) + \alpha_{+,3}(p_+ - p_{+,3})$, $\tilde{q}_h^- = \alpha_{-,2}(p_- - p_{-,2}) + \alpha_{-,3}(p_- - p_{-,3})$. The coefficients μ_+ and μ_- are computed as before by balancing the modified numerical fluxes

$$\tilde{q}_h^+\mu_+ + \tilde{q}_h^-\mu_- = 0$$

and using the convexity condition $\mu_+ + \mu_- = 1$. The regularized solution is

$$\mu_+ = \frac{|\tilde{q}_h^-| + \varepsilon}{|\tilde{q}_h^+| + |\tilde{q}_h^-| + 2\varepsilon}, \quad \mu_- = \frac{|\tilde{q}_h^+| + \varepsilon}{|\tilde{q}_h^+| + |\tilde{q}_h^-| + 2\varepsilon}. \tag{2.49}$$

For the case $\tilde{q}_h^+\tilde{q}_h^- > 0$, we obtain

$$q_h = (\mu_+\alpha_{+,1} + \mu_-\alpha_{-,1})(p_+ - p_-) = T_{+,1}(p_+ - p_-). \tag{2.50}$$

For the case $\tilde{q}_h^+\tilde{q}_h^- \le 0$, we get

$$\begin{aligned} q_h &= 2\mu_+\tilde{q}_h^+ + (\mu_+\alpha_{+,1} + \mu_-\alpha_{-,1})(p_+ - p_-) \\ &= -2\mu_-\tilde{q}_h^- - (\mu_+\alpha_{+,1} + \mu_-\alpha_{-,1})(p_+ - p_-), \end{aligned} \tag{2.51}$$

which for discretization (2.11) implies

$$\begin{aligned} q_h &= T_{+,2}(p_+ - p_{+,2}) + T_{+,3}(p_+ - p_{+,3}) + T_{+,1}(p_+ - p_-) \\ &= -T_{-,2}(p_- - p_{-,2}) - T_{-,3}(p_- - p_{-,3}) - T_{-,1}(p_- - p_+), \end{aligned} \tag{2.52}$$

where $T_{+,1}{=}T_{-,1}{=}\mu_+\alpha_{+,1}+\mu_-\alpha_{-,1}$. The coefficients $T_{+,1}$, $T_{+,2}$, $T_{+,3}$, $T_{-,1}$, $T_{-,2}$, and $T_{-,3}$ in (2.50), (2.52) are non-negative by construction and depend on the solution.

In all cases, the matrix $\mathbb{A}$ is assembled from matrices of flux discretization with zero row sum. Flux approximation at a boundary face is derived similar to the two-point flux discretization method. It may change the zero row sum to a positive row sum.

The approach to the construction of multi-point approximation in case of one-sided flux approximations (2.33) is similar. The two discrete fluxes (2.46) are expressed by

$$q_h^{\pm} = \pm c_{\pm,1} m_{\pm,1} \left(p_{\pm} - p_{\mp}\right) \pm c_{\pm,2} m_{\pm,2} \left(p_{\pm} - p_{\pm,2}\right) \pm c_{\pm,3} \left(m_{\pm,3} p_{\pm} - g_{\pm,3}\right), \tag{2.53}$$

in which we also identify the part of the stencil that excludes the two-point part:

$$\begin{aligned} q_h^{\pm} &= \pm c_{\pm,1} m_{\pm,1} \left(p_{\pm} - p_{\mp}\right) + \tilde{q}_h^{\pm}, \\ \tilde{q}_h^{\pm} &= \pm c_{\pm,2} m_{\pm,2} \left(p_{\pm} - p_{\pm,2}\right) \pm c_{\pm,3} \left(m_{\pm,3} p_{\pm} - g_{\pm,3}\right), \end{aligned} \tag{2.54}$$

and introduce $\mu_{\pm}$ according to (2.49). Then, the total flux has three expressions:

$$\begin{aligned} q_h &= \left(\mu_+ c_{+,1} m_{+,1} + \mu_- c_{-,1} m_{-,1}\right)\left(p_+ - p_-\right) + \left(\mu_+ \tilde{q}_h^+ - \mu_- \tilde{q}_h^-\right), \\ \hat{q}_h^+ &= \left(\mu_+ c_{+,1} m_{+,1} + \mu_- c_{-,1} m_{-,1}\right)\left(p_+ - p_-\right) + \begin{cases} 0, & \tilde{q}_h^+ \tilde{q}_h^- > 0 \\ 2\mu_+ \tilde{q}_h^+, & \tilde{q}_h^+ \tilde{q}_h^- \le 0 \end{cases}, \\ \hat{q}_h^- &= \left(\mu_+ c_{+,1} m_{+,1} + \mu_- c_{-,1} m_{-,1}\right)\left(p_+ - p_-\right) - \begin{cases} 0, & \tilde{q}_h^+ \tilde{q}_h^- > 0 \\ 2\mu_- \tilde{q}_h^-, & \tilde{q}_h^+ \tilde{q}_h^- \le 0 \end{cases}, \end{aligned} \tag{2.55}$$

where $\hat{q}_h^{\pm}$ as before yield definitions of positive coefficients $T_{\pm,i}$ for pressures and $T_{\pm,3,g}$ for boundary conditions. Here, $T_{+,1} = T_{-,1} = \left(\mu_+ c_{+,1} m_{+,1} + \mu_- c_{-,1} m_{-,1}\right)$, and the rest of coefficients are either $T_{\pm,2} = 2\mu_{\pm} c_{\pm,2} m_{\pm,2}$, $T_{\pm,3} = 2\mu_{\pm} c_{\pm,3} m_{\pm,2}$, $T_{\pm,3,g} = 2\mu_{\pm} c_{\pm,3}$ if $\tilde{q}_h^+ \tilde{q}_h^- \le 0$ or $T_{\pm,2} = T_{\pm,3} = T_{\pm,3,g} = 0$ otherwise.

There are two ways to assemble the matrix $\mathbb{A}$ with (2.55). The first way consists in using $\hat{q}_h^+$ and $\hat{q}_h^-$ when assembling (2.4) for ω_+ and ω_-, respectively. This results in the following contribution to the rows corresponding to ω_+ and ω_- in $\mathbb{A}$ and the right-hand side:

$$\begin{array}{l} \omega_+ \rightarrow \\ \omega_- \rightarrow \end{array} \left(\begin{array}{c|c|c|c} T_{+,1} + T_{+,2} + T_{+,3} & -T_{+,1} & -T_{+,2} & \\ -T_{-,1} & T_{-,1} + T_{-,2} + T_{-,3} & & -T_{-,2} \end{array} \right) \begin{bmatrix} p_+ \\ p_- \\ p_{+,2} \\ p_{-,2} \end{bmatrix} = \begin{bmatrix} T_{+,3,g} g_{+,3} \\ T_{-,3,g} g_{-,3} \end{bmatrix}. \tag{2.56}$$

It becomes evident that the row sums are non-negative in (2.56). As a result, the entire matrix $\mathbb{A}$ assembled from the internal fluxes has non-negative row sums. Let us further consider approximation at the boundary. We use one-sided approximation from (2.39):

$$q_h = q_h^+ = c_{+,1}m_{+,1}\left(p_+ - p_b\right) + c_{+,2}m_{+,2}\left(p_+ - p_{+,2}\right) + c_{+,3}\left(m_{+,3}p_+ - g_{+,3}\right). \tag{2.57}$$

In (2.57), the expression for q_h^+ already has non-negative row sum; however, p_b is not known in the cell-centered method. It is reconstructed from the boundary condition $\alpha p_b - \beta q_h = \gamma$ with non-negative parameters $\alpha \geq 0, \beta \geq 0$. Let us rewrite (2.57) for briefness of derivation as $q_h = \tilde{q}_h - c_{+,1}m_{+,1}p_b$; then, expressing p_b from the boundary condition yields

$$p_b = \left(\alpha + \beta c_{+,1}m_{+,1}\right)^{-1}\left(\gamma + \beta\tilde{q}_h\right), \tag{2.58}$$

and the boundary flux expression

$$\begin{aligned} q_h &= \left(\alpha + \beta c_{+,1}m_{+,1}\right)^{-1}\left(\alpha\tilde{q}^h - c_{+,1}m_{+,1}\gamma\right) \\ &= \left(\alpha + \beta c_{+,1}m_{+,1}\right)^{-1} c_{+,1}m_{+,1}\left(\alpha p_+ - \gamma\right) \\ &+ \left(\alpha + \beta c_{+,1}m_{+,1}\right)^{-1} \alpha c_{+,2}m_{+,2}\left(p_+ - p_{+,2}\right) \\ &+ \left(\alpha + \beta c_{+,1}m_{+,1}\right)^{-1} \alpha c_{+,3}\left(m_{+,3}p_+ - g_{+,3}\right), \end{aligned} \tag{2.59}$$

with non-negative coefficients

$$\begin{aligned} T_{+,1} &= \left(\alpha + \beta c_{+,1}m_{+,1}\right)^{-1} \alpha c_{+,1}m_{+,1}, \\ T_{+,2} &= \left(\alpha + \beta c_{+,1}m_{+,1}\right)^{-1} \alpha c_{+,2}m_{+,2}, \\ T_{+,3} &= \left(\alpha + \beta c_{+,1}m_{+,1}\right)^{-1} \alpha c_{+,3}m_{+,3}, \\ T_{+,1,\gamma} &= \left(\alpha + \beta c_{+,1}m_{+,1}\right)^{-1} c_{+,1}m_{+,1}, \\ T_{+,3,g} &= \left(\alpha + \beta c_{+,1}m_{+,1}\right)^{-1} \alpha c_{+,3}. \end{aligned} \tag{2.60}$$

The contribution of ω_+ to the row of $\mathbb{A}$ and the right-hand side is

$$\omega_+ \to \left(T_{+,1} + T_{+,2} + T_{+,3} \middle| -T_{+,2}\right)\begin{bmatrix} p_+ \\ p_{+,2} \end{bmatrix} = \left[T_{+,1,\gamma}\gamma + T_{+,3,g}g_{+,3}\right]. \tag{2.61}$$

The entire matrix $\mathbb{A}$ is assembled by blocks from (2.56) and (2.61) and has non-negative row sums. On each Picard iteration, the solution to the system involving $\mathbb{A}$ satisfies the discrete maximum principle [140, 148]. However, due to the difference in representation of each flux in the system, the discrete solution iterate p_h^k is not locally conservative until the nonlinear iterations converge.

Another way to assemble $\mathbb{A}$ is to use a single internal flux expression q_h from (2.55) in (2.4) to compute the contribution to the divergence for both cells ω_+ and ω_-. In this case each nonlinear iterate p_h^k is locally conservative but may violate the discrete maximum principle.

Converged solution satisfies both the discrete maximum principle and the conservation property in both approaches to $\mathbb{A}$ assembly.

2.5 Generalizations to Convection–Diffusion Equation

The above-discussed discretization methods are easily extended to a wider class of PDEs, the convection–diffusion equations:

$$\begin{aligned} \operatorname{div}(\mathbf{v}p - \mathbb{K}\nabla p) &= g, \text{ in } \Omega, \\ \alpha p + \beta \mathbf{n} \cdot (\mathbb{K}\nabla p) &= \gamma, \text{ on } \partial\Omega, \end{aligned} \tag{2.62}$$

where $\mathbf{v}(\mathbf{x})$ is a velocity field, $\mathbf{v} \in (L^\infty(\Omega))^3$, $\operatorname{div} \mathbf{v} \in L^\infty(\Omega)$, $\operatorname{div} \mathbf{v} \geq 0$ for almost every $\mathbf{x} \in \Omega$.

We denote by Γ_{out} the outflow part of $\partial\Omega$ where $\mathbf{v} \cdot \mathbf{n} \geq 0$, and define $\Gamma_{in} = \partial\Omega \setminus \Gamma_{out}$. Also, we assume that $\Gamma_N \subset \Gamma_{out}$. The set of boundary faces belonging to Γ_{in} and Γ_{out} is denoted by $\mathcal{F}_{in}$ and $\mathcal{F}_{out}$, respectively.

The difference between Eqs. (2.1) and (2.62) is the addition of the convective flux $\mathbf{v}p$ to the diffusion flux $-\mathbb{K}\nabla p$. Within the FV framework, a discretization of the convective flux on cell faces should be added to a discretization of the diffusion flux. The discretization of the total flux will have the same formulation (2.3) with stencil $\mathcal{P}_f$.

The simplest discretization of the averaged convective flux density $\frac{1}{|f|} \int_f p\mathbf{v} \cdot \mathbf{n}_f \, ds$ (further referred to as the convective flux, for brevity) on a mesh face f shared by two cells ω_+ and ω_- is the first-order upwind approximation

$$q_{h,\mathbf{v}} = v_f^+ \, p_+ + v_f^- \, p_-, \tag{2.63}$$

where

$$v_f^+ = \frac{1}{2}(v_f + |v_f|), \qquad v_f^- = \frac{1}{2}(v_f - |v_f|), \qquad v_f = \frac{1}{|f|} \int_f \mathbf{v} \cdot \mathbf{n}_f \, ds. \tag{2.64}$$

This linear discretization has the minimal stencil consisting of one cell and does not extend the stencil of the diffusive flux discretizations (2.3). However, it produces excessive numerical dissipation smearing out internal and boundary layers which are typical for singularly perturbed convection–diffusion equations.

Below we define the second-order upwind approximation of the convective flux which produces much lower numerical dissipation. Let a discontinuous piecewise linear function $\mathcal{R}(\mathbf{x})$ be defined via its restrictions $\mathcal{R}_\omega(\mathbf{x})$ to cells $\omega \in \Omega_h$. Then,

$$q_{h,\mathbf{v}} = v_f^+ \, \mathcal{R}_{\omega_+}(\mathbf{x}_f) + v_f^- \, \mathcal{R}_{\omega_-}(\mathbf{x}_f), \tag{2.65}$$

where $\mathbf{x}_f$ is the center of face f.

The function $\mathcal{R}(\mathbf{x})$ is a linear reconstruction of the discrete solution p_ω collocated at cell centers $\mathbf{x}_\omega$. The reconstruction has to be limited to get a monotone scheme:

$$\mathcal{R}_\omega(\mathbf{x}) = p_\omega + \mathcal{L}_\omega(\mathbf{g}_\omega) \cdot (\mathbf{x} - \mathbf{x}_\omega), \qquad \forall \mathbf{x} \in \omega, \tag{2.66}$$

where $\mathbf{g}_\omega$ is the reconstructed gradient and $\mathcal{L}_\omega$ is a limiting 3×3 matrix. The reconstruction (2.66) preserves the mean value of $\mathcal{R}_\omega(\mathbf{x})$ for any choice of $\mathcal{L}_\omega$. The gradient $\mathbf{g}$ minimizes on each cell ω the deviation functional

$$\mathcal{J}(\mathbf{g}_\omega) = \frac{1}{2} \sum_{\mathbf{x}_*} \left[p_\omega + \mathbf{g}_\omega \cdot (\mathbf{x}_* - \mathbf{x}_\omega) - p_* \right]^2,$$

where the summation is taken over $\mathbf{x}_*$, which are either the collocation points $\mathbf{x}_{\omega'}$ of the closest neighboring cells ω' or collocation points $\mathbf{x}_f$ of boundary faces $f \in \mathcal{F}_B \cap \mathcal{F}(\omega)$; $p_* = p_h(\mathbf{x}_*)$ is the value at the neighboring collocation point or boundary data at $\mathbf{x}_*$. The admissible gradient $\tilde{\mathbf{g}}_\omega = \mathcal{L}_\omega(\mathbf{g}_\omega)$ is such that linear reconstruction $\mathcal{R}$ is bounded at $\mathbf{x}_*$:

$$\min_{f \in \mathcal{F}_B \cap \mathcal{F}(\omega)} \left\{ p_{\omega'};\, p_f \right\} \le p_\omega + \tilde{\mathbf{g}}_\omega \cdot (\mathbf{x}_* - \mathbf{x}_\omega) \le \min_{f \in \mathcal{F}_B \cap \mathcal{F}(\omega)} \left\{ p_{\omega'};\, p_f \right\}, \quad \forall \mathbf{x}_*. \tag{2.67}$$

Due to (2.67), one has $\tilde{\mathbf{g}}_\omega \equiv 0$ in local minima and maxima. Note that according to the numerical evidence, for a singularly perturbed convection–diffusion equation the limiting (2.67) should be modified: the upper bound should involve only inflow boundary faces from $\mathcal{F}_{in}$.

$$\min_{f \in \mathcal{F}_B \cap \mathcal{F}(\omega)} \left\{ p_{\omega'};\, p_f \right\} \le p_\omega + \tilde{\mathbf{g}}_\omega \cdot (\mathbf{x}_* - \mathbf{x}_\omega) \le \min_{f \in \mathcal{F}_B \cap \mathcal{F}(\omega)} \left\{ p_{\omega'};\, p_{f_{in}} \right\},$$
$$\forall \mathbf{x}_* \in \{\mathbf{x}_{\omega'}, \mathbf{x}_f\}. \tag{2.68}$$

If one is interested in a positive scheme, the reconstructed function must satisfy the following restrictions at points $\mathbf{x}_f$ on faces f, where $v_f > 0$:

$$p_\omega + \tilde{\mathbf{g}}_\omega \cdot (\mathbf{x}_f - \mathbf{x}_\omega) \ge 0, \tag{2.69}$$

which guarantees correct sign of the advective flux. Normally, this condition follows from (2.67); however, if the face center $\mathbf{x}_f$ lies outside of the convex hull of points $\mathbf{x}_*$, the reconstructed function may become negative at this point.

Using (2.66), we represent the advective flux as the sum of a linear part (the first-order approximation) and a nonlinear part (the second-order correction):

$$q_{h,\mathbf{v}} = C_+ p_+ - C_- p_-, \tag{2.70}$$

where

$$C_\pm = \pm v_f^\pm (1 + \tilde{\mathbf{g}}_\pm \cdot (\mathbf{x}_f - \mathbf{x}_\pm) p_\pm^{-1}). \tag{2.71}$$

The coefficients $C_\pm$ are non-negative for positive $p_\pm$. If $p_\omega \geq 0$ for all cells $\omega \in \Omega_h$ and $p_{\omega'} = 0$ in a cell ω', then $\tilde{\mathbf{g}}_{\omega'}$ must be the zero vector and $C_\pm = \pm v_f^\pm$. Therefore, if the approximate FV solution of (2.62) is non-negative, the coefficients $C_\pm$ are non-negative.

If the solution of problem (2.62) may be negative in a part of Ω, then its FV approximation may be negative as well, and the sign of $C_\pm$ in (2.71) is unknown. However, using (2.70)–(2.71) in combination with the discrete diffusive fluxes (2.47), (2.50), (2.52) still provides the discrete minimum and maximum principles (2.109), (2.110).

For a boundary face, the approximation of the convective flux is applied for $f \in \mathcal{F}_D$ and depends on velocity direction on face f. If $f \in \mathcal{F}_{out}$, the approximation adopts formula (2.71). If $f \in \mathcal{F}_{in}$, we use

$$q_{h,\mathbf{v}} = -C_-, \tag{2.72}$$

where

$$C_- = -\frac{v_f^-}{|f|} \int_f \gamma \, \mathrm{d}s \geq 0. \tag{2.73}$$

The definition of the discrete convective flux for an internal face results in the addition to $\mathbb{T}_f$ of 2×2 matrices $\mathbb{C}_f$

$$\mathbb{C}_f = \begin{pmatrix} C_+ & -C_- \\ -C_+ & C_- \end{pmatrix}, \tag{2.74}$$

where the first column is zero if $v_f^+ = 0$ or the second column is zero if $v_f^- = 0$. For a face $f \in \mathcal{F}_{out}$, matrix $\mathbb{C}_f$ becomes 1×1 matrix. For a face $f \in \mathcal{F}_{in}$, the boundary data are assembled to the right-hand side only. The assembling procedure for the global convection–diffusion matrix $\mathbb{A}$ is similar to the diffusion case, $\mathbb{A}_{ij} := \mathbb{A}_{ij} + [\mathbb{D}_f(\mathbb{T}_f + \mathbb{C}_f)]_\pm$.

Although nonlinear discretization (2.65)–(2.66) has the minimal stencil consisting of one cell, its Jacobian has larger stencil than that for the diffusive flux discretizations (2.3): now it involves all cells ω' sharing a face with cell ω.

2.6 Generalization to Diffusion Problem in Mixed Formulation

The concept of a scalar flux on a cell face may be generalized to a concept of flux vectors which appears in a holistic approach to finite volume discretization of a system of PDEs with several unknown fields. Such approach collocates a degree of freedom of each field at the cell centers and derives approximation of the flux vector between two cells ω_+ and ω_- as a linear combination of unknown vectors collocated at barycenters $\mathbf{x}_{\omega_\pm}$ with matrix coefficients. In order to provide stability of the discretization, these matrix coefficients should have non-negative eigenvalues. We illustrate this concept on the simplest example of the diffusion problem (2.1) in the mixed formulation:

$$-\begin{bmatrix} & \nabla \\ \nabla^T \mathbb{K} & 0 \end{bmatrix} \begin{bmatrix} \mathbf{g} \\ p \end{bmatrix} = \begin{bmatrix} -\mathbf{g} \\ g \end{bmatrix}, \quad \text{in } \Omega, \tag{2.75}$$
$$\alpha p + \beta \mathbf{n} \cdot (\mathbb{K}\mathbf{g}) = \gamma, \qquad \text{on } \partial\Omega.$$

Here $\mathbf{g}$ is the gradient of unknown field p.

Integration of (2.76) over a cell ω gives

$$\int_{\partial\omega} \begin{bmatrix} & \mathbf{n} \\ \mathbf{n}^T \mathbb{K} & 0 \end{bmatrix} \begin{bmatrix} \mathbf{g} \\ p \end{bmatrix} \mathrm{d}s = \int_{\omega} \begin{bmatrix} \mathbf{g} \\ -g \end{bmatrix} \mathrm{d}\omega. \tag{2.76}$$

Denote by p_f and $\mathbf{g}_f$ averaged p and its gradient on face f, by $\mathbf{g}_\omega$ and g_ω averaged gradient and the source term on cell ω. Then, (2.76) becomes

$$\sum_{f \in \mathcal{F}(\omega)} \begin{bmatrix} & \mathbf{n}_f \\ \mathbf{n}_f^T \mathbb{K} & 0 \end{bmatrix} \begin{bmatrix} \mathbf{g}_f \\ p_f \end{bmatrix} |f| = \begin{bmatrix} \mathbf{g}_\omega \\ -g_\omega \end{bmatrix} |\omega|. \tag{2.77}$$

Formulation (2.77) implies the balance of the flux vectors $\mathbf{F}$

$$\int_{f \in \mathcal{F}(\omega)} \mathbf{F}|f| = \mathbf{G}|\omega|, \quad \mathbf{G} := \begin{bmatrix} \mathbf{g}_\omega \\ -g_\omega \end{bmatrix}, \tag{2.78}$$

where $\mathbf{F}$ is defined via the matrix coefficient A:

$$\mathbf{F} = A \begin{bmatrix} \mathbf{g}_f \\ p_f \end{bmatrix}, \qquad A = \begin{bmatrix} & \mathbf{n}_f \\ \mathbf{n}_f^T \mathbb{K} & 0 \end{bmatrix}. \tag{2.79}$$

Hereinafter we shall use the normal font for matrix coefficients associated with flux. Negative eigenvalue of A may cause instability of any discretization of (2.79) since A may produce the wrong direction of the flux. To suppress the instability, we split matrix A into a sum of singular matrices

$$A = A^> + A^<, \tag{2.80}$$

where $A^>$ has one positive eigenvalue

$$A^> = N^> \begin{bmatrix} \mathbb{K} & \\ & 1 \end{bmatrix}, \quad N^> \equiv \frac{1}{2} \begin{bmatrix} m^{-1}\mathbf{n}_f \mathbf{n}_f^T & \mathbf{n}_f \\ \mathbf{n}_f^T & m \end{bmatrix} = \frac{1}{2} \begin{bmatrix} m^{-1}\mathbf{n}_f \\ 1 \end{bmatrix} \begin{bmatrix} \mathbf{n}_f^T & m \end{bmatrix} \quad (2.81)$$

and $A^<$ has one negative eigenvalue

$$A^< = N^< \begin{bmatrix} \mathbb{K} & \\ & 1 \end{bmatrix}, \quad N^< \equiv \frac{1}{2} \begin{bmatrix} -m^{-1}\mathbf{n}_f \mathbf{n}_f^T & \mathbf{n}_f \\ \mathbf{n}_f^T & -m \end{bmatrix} = \frac{1}{2} \begin{bmatrix} -m^{-1}\mathbf{n}_f \\ 1 \end{bmatrix} \begin{bmatrix} \mathbf{n}_f^T & -m \end{bmatrix}. \quad (2.82)$$

Eigensplitting (2.80) holds for any positive parameter m. We shall fix it later.

We proceed to flux vector discretization on an interior face f shared by cells ω_+ and ω_- with diffusion tensors $\mathbb{K}_+$ and $\mathbb{K}_-$. Stable flux vector discretization uses matrix $A^<$ with a negative eigenvalue for the values collocated at cell ω_+ and matrix $A^>$ with a positive eigenvalue for the values collocated at cell ω_-. Let matrix coefficients $A_+ = A_+^> + A_+^<$ and $A_- = A_-^> + A_-^<$ be splittings (2.80), (2.81), (2.82) based on $\mathbb{K}_+$ and $\mathbb{K}_-$, respectively. In order to write one-sided first-order discretizations of flux vector (2.79), we denote by $\mathbf{g}_{f,\pm}$ the gradient of p on face f from the side of cell $\omega_\pm$. Then

$$A_+^< \left(\begin{bmatrix} \mathbf{g}_+ \\ p_+ \end{bmatrix} + O(h) \right) + A_+^> \begin{bmatrix} \mathbf{g}_{f,+} \\ p_f \end{bmatrix} = \mathbf{F} = A_-^> \left(\begin{bmatrix} \mathbf{g}_- \\ p_- \end{bmatrix} + O(h) \right) + A_-^< \begin{bmatrix} \mathbf{g}_{f,-} \\ p_f \end{bmatrix}. \quad (2.83)$$

Due to the divergence form of (2.1), $\mathbb{K}\nabla p$ belongs to the space $H(\text{div}, \Omega)$ and thus its normal component to all mesh faces is continuous, that is, $\mathbf{n}_f^T \mathbb{K}_+ \mathbf{g}_{f,+} = \mathbf{n}_f^T \mathbb{K}_- \mathbf{g}_{f,-}$. Due to factorizations of $N^>$ in (2.81) and $N^<$ in (2.82), only normal component of flux is involved in (2.83) and the face degrees of freedom can be eliminated from the continuity equation (2.83). Thus, the linear first-order flux discretization becomes

$$\mathbf{F} = N_+^> \left(N_+^> - N_-^<\right)^\dagger A_-^> \begin{bmatrix} \mathbf{g}_- \\ p_- \end{bmatrix} - N_-^< \left(N_+^> - N_-^<\right)^\dagger A_+^< \begin{bmatrix} \mathbf{g}_+ \\ p_+ \end{bmatrix} + O(h), \quad (2.84)$$

with pseudo-inverse matrix

$$\left(N_+^> - N_-^<\right)^\dagger = \left(\frac{m_+ + m_-}{2} \begin{bmatrix} (m_+ m_-)^{-1} \mathbf{n}_f \mathbf{n}_f^T & \mathbf{0} \\ \mathbf{0}^T & 1 \end{bmatrix} \right)^\dagger = \frac{2}{m_+ + m_-} \begin{bmatrix} m_+ m_- \mathbf{n}_f \mathbf{n}_f^T & \mathbf{0} \\ \mathbf{0}^T & 1 \end{bmatrix}, \quad (2.85)$$

for any positive m_+ and m_-.

Combination of (2.85) and (2.84) gives the final flux vector discretization:

$$\mathbf{F} = \frac{1}{m_+ + m_-} \left(\begin{bmatrix} \mathbf{n}_f \mathbf{n}_f^T \mathbb{K}_- & m_- \mathbf{n}_f \\ m_+ \mathbf{n}_f^T \mathbb{K}_- & m_+ m_- \end{bmatrix} \begin{bmatrix} \mathbf{g}_- \\ p_- \end{bmatrix} + \begin{bmatrix} -\mathbf{n}_f \mathbf{n}_f^T \mathbb{K}_+ & m_+ \mathbf{n}_f \\ m_- \mathbf{n}_f^T \mathbb{K}_+ & -m_+ m_- \end{bmatrix} \begin{bmatrix} \mathbf{g}_+ \\ p_+ \end{bmatrix} \right) + O(h). \quad (2.86)$$

The matrix coefficient for the degrees of freedom at cell $\omega_\pm$ has one non-zero eigenvalue $\lambda = \mp(\mathbf{n}_f^T\mathbb{K}_\pm\mathbf{n} + m_+m_-)/(m_+ + m_-)$ that corresponds to the general concept of flux vector discretization given in the beginning of this section.

Now we derive the discrete flux vector for a boundary face f belonging to cell ω_+ with permeability $\mathbb{K}_+$. The flux vector on f is approximated with the first order of accuracy by

$$\mathbf{F} = \widetilde{A}^> \begin{bmatrix} \mathbf{g}_f \\ p_f \end{bmatrix} + \widetilde{A}^< \left(\begin{bmatrix} \mathbf{g}_+ \\ p_+ \end{bmatrix} + O(h) \right). \tag{2.87}$$

The splitting of $A = \widetilde{A}^> + \widetilde{A}^<$ is such that the degrees of freedom $\mathbf{g}_f$ and p_f may be replaced by the boundary condition in (2.1)

$$\left[\, \beta\mathbf{n}_f^T\mathbb{K}_+ \;\; \alpha \,\right] \begin{bmatrix} \mathbf{g}_f \\ p_f \end{bmatrix} = \gamma_f, \tag{2.88}$$

where $\gamma_f = \dfrac{1}{|f|}\displaystyle\int_f \gamma(\mathbf{x})\mathrm{d}s$. The splitting which accounts (2.88) is

$$\widetilde{A}^> = \frac{1}{\alpha + \beta m_+} \begin{bmatrix} \mathbf{n}_f \\ m_+ \end{bmatrix} \left[\, \beta\mathbf{n}_f^T\mathbb{K}_+ \;\; \alpha \,\right], \quad \widetilde{A}^< = \frac{1}{\alpha + \beta m_+} \begin{bmatrix} -\beta\mathbf{n}_f\mathbf{n}_f^T\mathbb{K}_+ & \beta m_+\mathbf{n}_f \\ \alpha\mathbf{n}_f^T\mathbb{K}_+ & -\alpha m_+ \end{bmatrix}. \tag{2.89}$$

Note that multiplication by matrix $\widetilde{A}^>$ in (2.89) requires only the boundary condition (2.88). Matrix $\widetilde{A}^<$ has one non-zero eigenvalue $\lambda = \frac{\beta + \alpha m_+}{\alpha + \beta m_+}$. The final expression for the first-order approximation of the flux vector (2.87) is

$$\mathbf{F} = \frac{1}{\alpha + \beta m_+} \left(\begin{bmatrix} \mathbf{n}_f \\ m_+ \end{bmatrix} \gamma_f + \begin{bmatrix} -\beta\mathbf{n}_f\mathbf{n}_f^T\mathbb{K}_+ & \beta m_+\mathbf{n}_f \\ \alpha\mathbf{n}_f^T\mathbb{K}_+ & -\alpha m_+ \end{bmatrix} \begin{bmatrix} \mathbf{g}_+ \\ p_+ \end{bmatrix} \right) + O(h). \tag{2.90}$$

It remains to fix the parameter $m_\pm$. Numerically, small parameters $m_\pm$ produce non-monotone solutions, and large parameters $m_\pm$ reduce mesh convergence rate. A feasible compromise is provided by $m_\pm = \left(\mathbf{n}_f^T\mathbb{K}_\pm\mathbf{n}_f\right)^{3/2}$ for interior and boundary faces.

As reported in [144], the method is stable despite the collocation of both p and $\mathbf{g}$ at cell centers and manifestates the first-order convergence on numerous benchmarks as well as good monotonicity property.

2.7 Generalization to Navier–Stokes Equations

The dynamics of incompressible fluid with density ρ and dynamic viscosity μ is described by the incompressible Navier–Stokes equations in the flow domain Ω:

Table 2.1 Coefficients for different types of boundary conditions

	Γ_{Dir}	Γ_{noslip}	Γ_{slip}	$\Gamma_{tr-free}$	Γ_{MaxNav}	Γ_{pres}	$\Gamma_{do-noth}$
$\alpha_\perp$	1	1	1	0	1	0	0
$\alpha_\parallel$	1	1	0	0	λ	0	0
$\beta_\perp$	0	0	0	1	0	1	1
$\beta_\parallel$	0	0	1	1	1	1	1
$r_\perp$	$\mathbf{n}^T\mathbf{u}_b$	0	0	0	0	$-p_b$	0
$\mathbf{r}_\parallel$	$\left(\mathbb{I}-\mathbf{n}\mathbf{n}^T\right)\mathbf{u}_b$	$\mathbf{0}$	$\mathbf{0}$	$\mathbf{0}$	$\mathbf{0}$	$\mathbf{0}$	$\mathbf{0}$

$$\begin{aligned}\frac{\partial \rho \mathbf{u}}{\partial t} + \operatorname{div}\left(\rho \mathbf{u}\mathbf{u}^T - \boldsymbol{\tau}(\mathbf{u}, p)\right) &= \mathbf{f}, \\ \operatorname{div}(\mathbf{u}) &= 0.\end{aligned} \tag{2.91}$$

Here $\mathbf{u} = [u, v, w]^T$ is the velocity vector, p is the pressure, $\boldsymbol{\tau}(\mathbf{u}, p) = \mu\nabla\mathbf{u} - p\mathbb{I}$ is the stress tensor, and $\mathbf{f}$ is the body force (e.g., gravity).

The variety of boundary conditions involving velocity $\mathbf{u}_b$ and pressure p_b at the boundary with the unit normal $\mathbf{n}$ and representing different physical phenomena is given by equations

$$\begin{cases} \mathbf{n}^T\left(\alpha_\perp \mathbf{u}_b + \beta_\perp \boldsymbol{\tau}(\mathbf{u}, p_b)\mathbf{n}\right) = r_\perp, \\ \left(\mathbb{I} - \mathbf{n}\mathbf{n}^T\right)\left(\alpha_\parallel \mathbf{u}_b + \beta_\parallel \boldsymbol{\tau}(\mathbf{u}, p_b)\mathbf{n}\right) = \mathbf{r}_\parallel. \end{cases} \tag{2.92}$$

Table 2.1 combines typical variants of boundary conditions in terms of coefficients in (2.92). These boundary conditions are Dirichlet Γ_{Dir} (prescribed profile at the inlet), no-slip Γ_{noslip}, slip Γ_{slip}, traction-free $\Gamma_{tr-free}$, Maxwell–Navier Γ_{MaxNav}, prescribed outer pressure Γ_{pres} (balancing the normal stress), and natural do-nothing $\Gamma_{do-noth}$ (free outflow). In general, coefficients $\alpha_\perp, \beta_\perp, \alpha_\parallel, \beta_\parallel$ may be represented by tensors, e.g., tensor coefficient $\beta_\parallel$ may provide anisotropic surface friction.

Integration of (2.91) over a cell ω gives

$$\begin{aligned}\int_\omega \operatorname{div}\left(\rho \mathbf{u}\mathbf{u}^T - \boldsymbol{\tau}(\mathbf{u}, p)\right) \mathrm{d}\omega &= \int_{\partial\omega} \left(\rho \mathbf{u}\mathbf{u}^T - \boldsymbol{\tau}(\mathbf{u}, p)\right) \mathrm{d}\mathbf{s} \\ &= \sum_{f \in \mathcal{F}(\omega)} |f| \left[\rho \mathbf{u}\mathbf{u}^T\mathbf{n}_f - \boldsymbol{\tau}(\mathbf{u}, p)\mathbf{n}_f\right]_{\mathbf{x}_f} + O(h),\end{aligned}$$

$$\int_\omega \operatorname{div}(\mathbf{u})\, \mathrm{d}\omega = \int_{\partial\omega} \mathbf{u} \cdot \mathrm{d}\mathbf{s} = \sum_{f \in \mathcal{F}(\omega)} |f| \left[\mathbf{u}^T\mathbf{n}_f\right]_{\mathbf{x}_f} + O(h),$$

$$\int_\omega \frac{\partial \rho \mathbf{u}}{\partial t} \mathrm{d}\omega = |\omega| \rho \left[\frac{\partial \mathbf{u}}{\partial t}\right]_{\mathbf{x}_\omega} + O(h),$$

for face center $\mathbf{x}_f \in f$ and cell center $\mathbf{x}_\omega \in \omega$. Replacing $\frac{\partial \mathbf{u}}{\partial t}$ by an implicit finite difference approximation in time (e.g., the first-order backward Euler) $\left(\frac{\partial \mathbf{u}}{\partial t}\right)^n$ at time step t^n, one arrives at the balance equation at time step n

$$\begin{aligned} \rho\left(\frac{\partial \mathbf{u}}{\partial t}\right)^n |\omega| + \sum_{f\in\mathcal{F}(\omega)} |f|\mathbf{t} &= \int_\omega \mathbf{f}(t^n)\mathrm{d}\omega \\ \sum_{f\in\mathcal{F}(\omega)} |f|q &= 0, \end{aligned} \tag{2.93}$$

where $\mathbf{t}$ and q denote momentum flux vector and continuity flux

$$\mathbf{t} = \mathbf{t}_a + \mathbf{t}_t + \mathbf{t}_p,\ \mathbf{t}_a = \rho\left[\mathbf{u}\mathbf{u}^T\mathbf{n}\right]_{\mathbf{x}_f},\ \mathbf{t}_t = -\mu\left[\nabla\mathbf{u}\mathbf{n}\right]_{\mathbf{x}_f},\ \mathbf{t}_p = [p\mathbf{n}]_{\mathbf{x}_f},\ q = \left[\mathbf{n}^T\mathbf{u}\right]_{\mathbf{x}_f}. \tag{2.94}$$

Here, advection flux vector $\mathbf{t}_a$, traction flux vector $\mathbf{t}_t$ and pressure flux vector $\mathbf{t}_p$ contribute to the momentum flux vector $\mathbf{t}$, and the velocity gradient is defined by

$$\nabla\mathbf{u} := \begin{bmatrix} \partial_x u & \partial_y u & \partial_z u \\ \partial_x v & \partial_y v & \partial_z v \\ \partial_x w & \partial_y w & \partial_z w \end{bmatrix}.$$

We follow the general approach (see Sect. 2.7) of deriving a stable discretization of flux vectors through matrix coefficients with non-negative eigenvalues. To define a cell-centered finite volume scheme, for any interior face f shared by two cells $\omega_\pm$ we consider discrete velocities $\mathbf{u}_\pm$ and pressures $p_\pm$ collocated at the cell centers $\mathbf{x}_\pm$, and assume for a moment that the discrete velocity $\mathbf{u}_f$ at the face center $\mathbf{x}_f$ is known.

First, from the Taylor expansion at point $\mathbf{x}_+$ and assumption of close-to-linear velocity behavior, we derive two approximations of $\mathbf{t}_a$:

$$\begin{aligned} \mathbf{t}_a &\approx \frac{1}{2}\rho Q(\mathbf{u}_+)\left(3\mathbf{u}_+ - 2\mathbf{u}_f + 4\left[\mathbb{I}\otimes(\mathbf{x}_f - \mathbf{x}_+)^T\right][\nabla\otimes\mathbf{u}_+]\right), \\ \mathbf{t}_a &\approx \frac{1}{2}\rho Q(\mathbf{u}_+)\left(2\mathbf{u}_f - \mathbf{u}_+\right), \end{aligned}$$

where $[\nabla\otimes\mathbf{u}_i] := [\nabla u_i, \nabla v_i, \nabla w_i]^T$, $Q(\mathbf{u}_+) := \mathbf{u}_+\mathbf{n}^T + \mathbf{n}^T\mathbf{u}_+\mathbb{I}$ is a 3×3 effective velocity tensor that has eigenvalues of the same sign as $\mathbf{n}^T\mathbf{u}_+$, and the magnitude of the largest eigenvalue is $2\left|\mathbf{n}^T\mathbf{u}_+\right|$. We recall that we use the normal font for matrix coefficients (*e.g.* Q) associated with flux.

Second, since $\mathbf{t}_t = -\mu\nabla\mathbf{u}\mathbf{n} = -\mu[\mathbb{I}\otimes\mathbf{n}^T][\nabla\otimes\mathbf{u}]$, we get

$$\mathbf{t}_t \approx \frac{\mu}{r_+}(\mathbf{u}_+ - \mathbf{u}_f) - \left(\mu[\mathbb{I}\otimes\mathbf{n}^T] - \frac{\mu}{r_+}\left[\mathbb{I}\otimes(\mathbf{x}_f - \mathbf{x}_+)^T\right]\right)[\nabla\otimes\mathbf{u}_+],$$

where $r_+ = \mathbf{n}^T\left(\mathbf{x}_f - \mathbf{x}_+\right)$ is the distance from $\mathbf{x}_+$ to f-plane.

Third,

$$\begin{bmatrix} \mathbf{t}_p \\ q \end{bmatrix} = \begin{bmatrix} & \mathbf{n} \\ \mathbf{n}^T & \end{bmatrix} \begin{bmatrix} \mathbf{u}_f \\ p_f \end{bmatrix} \approx \begin{bmatrix} & \mathbf{n} \\ \mathbf{0}^T & \end{bmatrix} \begin{bmatrix} \mathbf{u}_+ \\ p_+ \end{bmatrix} + \begin{bmatrix} & \mathbf{0} \\ \mathbf{n}^T & \end{bmatrix} \begin{bmatrix} \mathbf{u}_f \\ \end{bmatrix},$$

where the approximation is derived assuming close-to-constant pressure field in cell ω_+, $p_f = p_+$.

Now we can split the matrix coefficients in expressions for

$$\begin{aligned} \mathbf{t}_a &\approx \frac{1}{2}\rho Q^>(\mathbf{u}_+)\left(3\mathbf{u}_+ - 2\mathbf{u}_f + 4\left[\mathbb{I} \otimes \left(\mathbf{x}_f - \mathbf{x}_+\right)^T\right][\nabla \otimes \mathbf{u}_+]\right) \\ &+ \frac{1}{2}\rho Q^<(\mathbf{u}_+)\left(2\mathbf{u}_f - \mathbf{u}_+\right), \end{aligned} \tag{2.95}$$

and

$$\begin{aligned} \begin{bmatrix} \mathbf{t}_p \\ q \end{bmatrix} &\approx \begin{bmatrix} \xi_+\mathbf{n}\mathbf{n}^T & \mathbf{n} \\ \mathbf{0}^T & \end{bmatrix} \begin{bmatrix} \mathbf{u}_+ \\ p_+ \end{bmatrix} + \begin{bmatrix} -\xi_+\mathbf{n}\mathbf{n}^T & \mathbf{0} \\ \mathbf{n}^T & \end{bmatrix} \begin{bmatrix} \mathbf{u}_f \\ \end{bmatrix} \\ &+ \begin{bmatrix} \xi_+\left[\mathbf{n}\mathbf{n}^T \otimes (\mathbf{x}_f - \mathbf{x}_+)^T\right][\nabla \otimes \mathbf{u}_+] \\ \end{bmatrix}, \end{aligned} \tag{2.96}$$

where $Q(\mathbf{u}_+) = Q^>(\mathbf{u}_+) + Q^<(\mathbf{u}_+)$, $Q^>(\mathbf{u}_+) - Q^<(\mathbf{u}_+) = \left|\mathbf{n}^T\mathbf{u}_+\right|_\varepsilon \left(\mathbb{I} + \mathbf{n}\mathbf{n}^T\right)$, $|a|_\varepsilon := \sqrt{a^2 + \varepsilon^2}$, ξ_+ is a positive parameter to be chosen later.

Based on these splittings, we rewrite two one-sided momentum flux approximations:

$$\begin{aligned} \mathbf{t} &\approx \left(T_+ + \frac{1}{2}Q_+\right)\mathbf{u}_+ - T_+\mathbf{u}_f + p_+\mathbf{n} \\ &+ \left((T_+ + Q_+)\left[\mathbb{I} \otimes (\mathbf{x}_f - \mathbf{x}_+)^T\right] - \mu\left[\mathbb{I} \otimes \mathbf{n}^T\right]\right)[\nabla \otimes \mathbf{u}_+] \\ \mathbf{t} &\approx T_-\mathbf{u}_f - \left(T_- - \frac{1}{2}Q_-\right)\mathbf{u}_- + p_-\mathbf{n} \\ &+ \left((T_- - Q_-)\left[\mathbb{I} \otimes (\mathbf{x}_- - \mathbf{x}_f)^T\right] - \mu\left[\mathbb{I} \otimes \mathbf{n}^T\right]\right)[\nabla \otimes \mathbf{u}_-], \end{aligned} \tag{2.97}$$

with notations (for the sake of brevity)

$$T_\pm = \rho\left(Q^>(\mathbf{u}_\pm) - Q^<(\mathbf{u}_\pm)\right) + \frac{\mu}{r_\pm}\mathbb{I} + \xi_\pm\mathbf{n}\mathbf{n}^T, \quad Q_\pm = \rho Q(\mathbf{u}_\pm).$$

Equating the two one-sided approximations, we eliminate $\mathbf{u}_f$:

$$\begin{aligned} \mathbf{u}_f &= (T_+ + T_-)^{-1}\left(\left(T_+ + \frac{1}{2}Q_+\right)\mathbf{u}_+ + \left(T_- - \frac{1}{2}Q_-\right)\mathbf{u}_- + (p_+ - p_-)\mathbf{n}\right) \\ &+ (T_+ + T_-)^{-1}\left((T_+ + Q_+)\left[\mathbb{I} \otimes (\mathbf{x}_f - \mathbf{x}_+)^T\right] - \mu\left[\mathbb{I} \otimes \mathbf{n}^T\right]\right)[\nabla \otimes \mathbf{u}_+] \\ &- (T_+ + T_-)^{-1}\left((T_- - Q_-)\left[\mathbb{I} \otimes (\mathbf{x}_- - \mathbf{x}_f)^T\right] - \mu\left[\mathbb{I} \otimes \mathbf{n}^T\right]\right)[\nabla \otimes \mathbf{u}_-] \end{aligned} \tag{2.98}$$

and get approximation of $\mathbf{t}$:

$$\begin{aligned}\mathbf{t} \approx\ & T_-(T_+ + T_-)^{-1}\left(\left(T_+ + \frac{1}{2}Q_+\right)\mathbf{u}_+ + p_+\mathbf{n}\right)\\ & - T_+(T_+ + T_-)^{-1}\left(\left(T_- - \frac{1}{2}Q_-\right)\mathbf{u}_- - p_-\mathbf{n}\right)\\ & + T_-(T_+ + T_-)^{-1}\left((T_+ + Q_+)\left[\mathbb{I}\otimes(\mathbf{x}_f - \mathbf{x}_+)^T\right] - \mu\left[\mathbb{I}\otimes\mathbf{n}^T\right]\right)\left[\nabla\otimes\mathbf{u}_+\right]\\ & + T_+(T_+ + T_-)^{-1}\left((T_- - Q_-)\left[\mathbb{I}\otimes(\mathbf{x}_- - \mathbf{x}_f)^T\right] - \mu\left[\mathbb{I}\otimes\mathbf{n}^T\right]\right)\left[\nabla\otimes\mathbf{u}_-\right].\end{aligned} \tag{2.99}$$

To summarize, for interior face f, the flux vector discretization at $\mathbf{x}_f$ becomes

$$\begin{aligned}\begin{bmatrix}\mathbf{t}\\ q\end{bmatrix} \approx & \begin{bmatrix}T_-(T_+ + T_-)^{-1} & \\ & 1\end{bmatrix}\begin{bmatrix}T_+ + T_- & \mathbf{n}\\ \mathbf{n}^T & \mathbf{n}^T(T_+ + T_-)^{-1}\mathbf{n}\end{bmatrix}\begin{bmatrix}(T_+ + T_-)^{-1}(T_+ + \frac{1}{2}Q_+) & \\ & 1\end{bmatrix}\begin{bmatrix}\mathbf{u}_+\\ p_+\end{bmatrix}\\ & - \begin{bmatrix}T_+(T_+ + T_-)^{-1} & \\ & 1\end{bmatrix}\begin{bmatrix}T_+ + T_- & -\mathbf{n}\\ -\mathbf{n}^T & \mathbf{n}^T(T_+ + T_-)^{-1}\mathbf{n}\end{bmatrix}\begin{bmatrix}(T_+ + T_-)^{-1}(T_- - \frac{1}{2}Q_-) & \\ & 1\end{bmatrix}\begin{bmatrix}\mathbf{u}_-\\ p_-\end{bmatrix}\\ & + \begin{bmatrix}T_-\\ \mathbf{n}^T\end{bmatrix}(T_+ + T_-)^{-1}\left((T_+ + Q_+)\left[\mathbb{I}\otimes(\mathbf{x}_f - \mathbf{x}_+)^T\right] - \mu\left[\mathbb{I}\otimes\mathbf{n}^T\right]\right)\left[\nabla\otimes\mathbf{u}_+\right]\\ & + \begin{bmatrix}T_+\\ -\mathbf{n}^T\end{bmatrix}(T_+ + T_-)^{-1}\left((T_- - Q_-)\left[\mathbb{I}\otimes(\mathbf{x}_- - \mathbf{x}_f)^T\right] - \mu\left[\mathbb{I}\otimes\mathbf{n}^T\right]\right)\left[\nabla\otimes\mathbf{u}_-\right].\end{aligned} \tag{2.100}$$

The discretization (2.100) is inf-sup stable since matrices

$$\begin{bmatrix}T_+ + T_- & \pm\mathbf{n}\\ \pm\mathbf{n}^T & \mathbf{n}^T(T_+ + T_-)^{-1}\mathbf{n}\end{bmatrix}$$

have non-negative eigenvalues for any positive $\xi_\pm$ and matrices $T_\pm \pm Q_\pm$ have positive eigenvalues due to $Q^>(\mathbf{u}_\pm) - Q^<(\mathbf{u}_\pm) = |\mathbf{n}^T\mathbf{u}_\pm|_\varepsilon\left(\mathbb{I} + \mathbf{n}\mathbf{n}^T\right)$.

The stabilization parameter $\xi_\pm$ in (2.96) is used to suppress error in pressure:

$$\xi_\pm = \frac{l}{r_\pm}\sqrt{\mathbf{u}_\pm^T\left(\mathbb{I} - \mathbf{n}\mathbf{n}^T\right)\mathbf{u}_\pm}, \quad l = 6\alpha|\Omega|/|\partial\Omega|, \tag{2.101}$$

where l is characteristic domain length and α is a free scaling parameter, we take $\alpha \sim 10^{-1}$.

For a boundary face f belonging to cell ω_+, we rewrite (2.92) as

$$A_D\mathbf{u}_b + A_N\left(\mu\left[\mathbb{I}\otimes\mathbf{n}^T\right]\left[\nabla\otimes\mathbf{u}_1\right] - p_b\mathbf{n}\right) = A_R, \tag{2.102}$$

where

$$A_D := \alpha_\perp\mathbf{n}\mathbf{n}^T + \alpha_\parallel\left(\mathbb{I} - \mathbf{n}\mathbf{n}^T\right), \quad A_N := \beta_\perp\mathbf{n}\mathbf{n}^T + \beta_\parallel\left(\mathbb{I} - \mathbf{n}\mathbf{n}^T\right), \quad A_R := \mathbf{n}r_\perp + \mathbf{r}_\parallel.$$

Now, assuming close-to-constant pressure field in cell ω_+, we set $p_b = p_+$ and use the approximation of the vector traction flux $\mathbf{t}_t$ to write at point $\mathbf{x}_f$

$$
\begin{aligned}
&A_D \mathbf{u}_f + A_N \left(T_b(\mathbf{u}_f - \mathbf{u}_+) - p_+ \mathbf{n} \right) \\
&\quad + A_N \left(\mu \left[\mathbb{I} \otimes \mathbf{n}^T \right] - T_b \left[\mathbb{I} \otimes \left(\mathbf{x}_f - \mathbf{x}_+ \right)^T \right] \right) [\nabla \otimes \mathbf{u}_+] = A_R,
\end{aligned}
\tag{2.103}
$$

with a 3×3 matrix coefficient $T_b \equiv T_+$. From (2.103), we eliminate $\mathbf{u}_f$:

$$
\begin{aligned}
\mathbf{u}_f &= (A_D + A_N T_b)^{-1} \left(A_R + A_N \left(T_b \mathbf{u}_+ + p_+ \mathbf{n} \right) \right) \\
&\quad + (A_D + A_N T_b)^{-1} \left(T_b \left[\mathbb{I} \otimes \left(\mathbf{x}_f - \mathbf{x}_+ \right)^T \right] - \mu \left[\mathbb{I} \otimes \mathbf{n}^T \right] \right) [\nabla \otimes \mathbf{u}_+]
\end{aligned}
\tag{2.104}
$$

and the final discrete flux vector at $\mathbf{x}_f$ becomes

$$
\begin{aligned}
\begin{bmatrix} \mathbf{t} \\ q \end{bmatrix} \approx & \left[\begin{array}{c|c} T_+ (A_D + A_N T_b)^{-1} A_D + \frac{1}{2} Q_+ & T_+ (A_D + A_N T_b)^{-1} (A_D + A_N (T_b - T_+)) T_+^{-1} \mathbf{n} \\ \hline \mathbf{n}^T (A_D + A_N T_b)^{-1} A_N T_b & \mathbf{n}^T (A_D + A_N T_b)^{-1} A_N \mathbf{n} \end{array} \right] \begin{bmatrix} \mathbf{u}_+ \\ p_+ \end{bmatrix} \\
& - \begin{bmatrix} T_+ \\ -\mathbf{n}^T \end{bmatrix} (A_D + A_N T_b)^{-1} A_R \\
& - \begin{bmatrix} T_+ (A_D + A_N T_b)^{-1} (A_D + A_N (T_b - T_+)) T_+^{-1} \\ \mathbf{n}^T (A_D + A_N T_b)^{-1} A_N \end{bmatrix} \mu \left[\mathbb{I} \otimes \mathbf{n}^T \right] [\nabla \otimes \mathbf{u}_+] \\
& + \begin{bmatrix} T_+ (A_D + A_N T_b)^{-1} A_D + Q_+ \\ \mathbf{n}^T (A_D + A_N T_b)^{-1} A_N T_b \end{bmatrix} \left[\mathbb{I} \otimes (\mathbf{x}_\sigma - \mathbf{x}_+)^T \right] [\nabla \otimes \mathbf{u}_+].
\end{aligned}
\tag{2.105}
$$

It remains to recover the gradient $[\nabla \otimes \mathbf{u}_\pm]$ in cells $\omega_\pm$. Continuity of the velocity $\mathbf{u}$ implies for any interior face

$$
\left[\mathbb{I} \otimes (\mathbf{x}_- - \mathbf{x}_+)^T \right] [\nabla \otimes \mathbf{u}_+] = \mathbf{u}_- - \mathbf{u}_+,
\tag{2.106}
$$

for any boundary face

$$
\left(A_D \left[\mathbb{I} \otimes \left(\mathbf{x}_f - \mathbf{x}_+ \right)^T \right] + \mu A_N \left[\mathbb{I} \otimes \mathbf{n}^T \right] \right) [\nabla \otimes \mathbf{u}_+] = A_R - A_D \mathbf{u}_+ + p_+ A_N \mathbf{n},
\tag{2.107}
$$

due to $\left[\mathbb{I} \otimes \left(\mathbf{x}_f - \mathbf{x}_+ \right)^T \right] [\nabla \otimes \mathbf{u}_+] = \mathbf{u}_f - \mathbf{u}_+$.

Assembling the conditions (2.106) and (2.107) over all faces of ω_+ forms the system $A[\nabla \otimes \mathbf{u}_+] = R$, with matrix $A \in \Re^{3|\mathcal{F}(\omega_+)| \times 9}$ and right-hand side vector $R \in \Re^{3|\mathcal{F}(\omega_+)|}$. The overdetermined system is solved with the least squares method by $[\nabla \otimes \mathbf{u}_+] = \left(A^T A \right)^{-1} A^T R$ where the Cholesky method is employed to invert 9×9 positive-definite matrix $A^T A$.

Brief analysis demonstrates the second-order approximation of all terms in (2.91) except pressure whose contribution to the momentum flux gives the first order of accuracy. Numerical experiments prove that the convergence rate both for velocity and pressure may be lower than the second order although is always higher than the first order.

Although the presented FV scheme is based on a linear two-point flux approximation, the algebraic system for unknown $\mathbf{u}^n$, p^n at the nth time step is nonlinear due to nonlinearity of the advection term in (2.91). The solution of the nonlinear system is based on the Newton method. Denote the velocity and pressure of kth iteration of the Newton method by $\mathbf{u}^k$, p^k. The initial guess for the iterations is taken from the previous time step, $\mathbf{u}^{k=0} \equiv \mathbf{u}^{n-1}$, $p^{k=0} \equiv p^{n-1}$. The residual vector $\mathcal{R}^k$ of order $4|C(\Omega_h)|$ and the Jacobian matrix $\mathcal{J}^k$ of size $4|C(\Omega_h)| \times 4|C(\Omega_h)|$ are assembled over all mesh faces.

If the L_2-norm $\|\mathcal{R}^k\|_{L_2}$ of the residual satisfies convergence criteria, we declare $\mathbf{u}^{n+1} = \mathbf{u}^k$, $p^{n+1} = p^k$; otherwise, we solve the system for the Newton update

$$\mathcal{J}^k \left(\begin{bmatrix} \mathbf{u}^{k+1} \\ p^{k+1} \end{bmatrix} - \begin{bmatrix} \mathbf{u}^k \\ p^k \end{bmatrix} \right) = \mathcal{R}^k. \tag{2.108}$$

Implementation of the Newton method is easy within the INMOST platform. Given assembly procedure for the residual $\mathcal{R}^k$, INMOST provides automatic differentiation for the sparse Jacobian assembly and iterative linear solvers for systems with distributed sparse matrices, rf. to Chap. 6.

2.8 Analysis of Monotone FV Methods

Nonlinearity of the monotone discretizations complicates considerably their analysis. Even existence of the solution of (2.7) is the open question although some steps toward the existence have been performed. The convergence was proved under scheme coercivity assumption in [59, 131]. For numerical convergence of these schemes on general meshes, we refer to [48, 51, 105–107, 114, 115, 142–144]. In most cases, the second-order convergence for unknown p and the first-order convergence for flux q is observed. Keeping in mind applications, we confine ourselves by monotonicity properties of the schemes.

Differential solution of both (2.1) and (2.62) satisfies the maximum and minimum principles. For simplicity, we consider the case $\alpha = 1$, $\beta = 0$, i.e., $\Gamma_D = \partial\Omega$. The minimum principle states that for $g \geq 0$, $\operatorname{div}\mathbf{v} \geq 0$, the solution $p(\mathbf{x})$ satisfies [100]:

$$\min_{\mathbf{x}\in\bar{\Omega}} p(\mathbf{x}) \geq \min_{\mathbf{x}\in\partial\Omega} \gamma(\mathbf{x}).$$

The maximum principle is formulated accordingly: for $g \leq 0$, the solution $p(\mathbf{x})$ satisfies

$$\max_{\mathbf{x}\in\bar{\Omega}} p(\mathbf{x}) \leq \max_{\mathbf{x}\in\partial\Omega} \gamma(\mathbf{x}).$$

Henceforth, we shall refer to both principles as the maximum principle.

The direct corollary of the maximum principle is non-negativity of $p(\mathbf{x})$ for $g \geq 0$, $\gamma \geq 0$. If $g = 0$ then non-oscillatory boundary condition $\gamma(\mathbf{x})$ implies non-oscillatory solution $p(\mathbf{x})$.

2.8.1 Two-Point Flux Approximations

Monotonicity analysis of nonlinear two-point flux approximations (2.14) and (2.37) is based on non-negativity of coefficients T_+ and T_- which is guaranteed by construction. Application of the Picard iterations to the solution of (2.7) results in the solution of linear systems with matrix $\mathbb{A}(\mathbf{p}^k)$. The matrix is assembled from matrices $\mathbb{D}_f \mathbb{T}_f$ from (2.6). The matrix $\mathbb{A}(\mathbf{p}^k)$ has positive diagonal entries, non-negative column sums; it is irreducible due to the consistency of the polyhedral mesh, and therefore, it is an M-matrix that is known to give non-negative solution of systems with non-negative right-hand side [140, 148]. The right-hand side is non-negative by construction for $\mathbf{p} \geq 0$. If one starts from a non-negative initial guess $\mathbf{p}^0 \geq 0$, then all iterates $\mathbf{p}^k$ will be non-negative. If the Picard iterations converge, the solution of (2.7) will be non-negative as well.

Contribution of the discrete convective flux (2.63) or (2.65) does not change the properties of matrix $\mathbb{A}(\mathbf{p}^k)$, and thus the solution of (2.7) remains non-negative [107].

2.8.2 Multi-Point Flux Approximation

The nonlinear multi-point flux approximation (2.47), (2.50), (2.52) or (2.57), (2.59) is designed to satisfy (for $\Gamma_D = \partial\Omega$)

Discrete minimum and maximum principles. *Let a solution* $\mathbf{p}$ *to* (2.7) *exist.*

If $g \geq 0$, div$\mathbf{v} \geq 0$, then

$$\min_{\omega\in\Omega_h} p_\omega \geq p_{\min} \equiv \min_{f\in\mathcal{F}_B} p_f. \tag{2.109}$$

If $g \leq 0$, div$\mathbf{v} \geq 0$, then

$$\max_{\omega\in\Omega_h} p_\omega \leq p_{\max} \equiv \max_{f\in\mathcal{F}_B} p_f. \tag{2.110}$$

For brevity, we shall refer to both principles as the discrete maximum principle (DMP).

Monotonicity analysis for problem (2.1) is also based on algebraic results [32, 140]. Irreducibility of $\mathbb{A}(\mathbf{p}^k)$ is assumed to follow from the consistency of the polyhedral mesh. The matrix $\mathbb{A}(\mathbf{p}^k)$ has positive diagonal entries, non-negative row sums; it is assumed to be irreducible and therefore the Picard iterate $\mathbb{A}(\mathbf{p}^k)^{-1}\mathbf{p}^{k+1}$ satisfies the discrete maximum principle. If the Picard iterations converge, the solution to (2.7) satisfies the DMP (2.109), (2.110) as well.

For the convection–diffusion equation (2.62) with the first-order upwind approximation of convective fluxes (2.63), matrix $\mathbb{A}$ has diagonal dominance in rows: the diffusion part of the matrix $\mathbb{A}$ has non-negative row sum (see above), and the convective part of the matrix with entries c_{ij} has also non-negative row sum as shown below.

We distinguish two cases. If a cell ω is not adjacent to Γ_{in}, for the associated row i of the convective part we have

$$\sum_j c_{ij} = \sum_{f \in \mathcal{F}(\omega)} v_f^+ \cdot 1 + v_f^- \cdot 1 = \sum_{f \in \mathcal{F}(\omega)} v_f = \int_\omega \operatorname{div}\mathbf{v}\, d\omega \geq 0$$

due to (2.63), (2.64), and the assumption $\operatorname{div}\mathbf{v} \geq 0$. If a cell ω is adjacent to Γ_{in} in faces $\mathcal{F}_{in}(\omega)$, we have

$$\begin{aligned}
\sum_j c_{ij} &= \sum_{f \in \mathcal{F}(\omega)\setminus\mathcal{F}_{in}(\omega)} v_f^+ \cdot 1 + v_f^- \cdot 1 = \sum_{f \in \mathcal{F}(\omega)} v_f^+ + v_f^- - \sum_{f \in \mathcal{F}_{in}(\omega)} v_f^- \\
&= \int_\omega \operatorname{div}\mathbf{v}\, d\omega - \sum_{f \in \mathcal{F}_{in}(\omega)} v_f^- \geq \int_\omega \operatorname{div}\mathbf{v}\, d\omega \geq 0,
\end{aligned}$$

since $v_f^- \leq 0$. Therefore, due to the diagonal dominance in rows of $\mathbb{A}(\mathbf{p}^k)$ the same algebraic result [32, 140] can be applied to prove (2.109), (2.110) both for each Picard iterate and the solution of (2.7).

For the second-order upwind approximation (2.65), the proof of (2.109), (2.110) uses ellipticity of the convection–diffusion operator due to $\mathbb{K}(\mathbf{x}) > 0$, $\operatorname{div}\mathbf{v} \geq 0$ and the special multi-point form of the diffusive flux (2.47), (2.50), (2.52) whose assembled matrix is denoted by $\mathbb{A}_{dif}$. The latter is assumed to be an irreducible matrix. We shall prove (2.109) by a contradiction, and (2.110) is proven by analogy.

Let a cell ω have the smallest value p_ω and $p_\omega < p_{\min}$. Without loss of generality, we assume that vectors $\mathbf{n}_f$, $f \in \mathcal{F}(\omega)$, are exterior to ω.

Let ω'_f denote a cell sharing face f with the cell ω. Since p_ω is the global minimum, we get $\tilde{\mathbf{g}}_\omega = 0$ and $\mathcal{R}_\omega \equiv p_\omega$ due to (2.67), $\mathcal{R}_{\omega'_f}(\mathbf{x}_f) \geq p_\omega$ and $p_f > p_\omega$. Due to the definition of convective fluxes

$$\sum_{f \in \mathcal{F}(\omega)} q_{h,\mathbf{v}} = \sum_{f \in \mathcal{F}(\omega)\setminus\mathcal{F}_{in}(\omega)} \left(v_f^+ p_\omega + v_f^- \mathcal{R}_{\omega'_f}(\mathbf{x}_f)\right) + \sum_{f \in \mathcal{F}(\omega)\cap\mathcal{F}_{in}(\omega)} v_f^- p_f.$$

Using $v_f^- = v_f$ on inflow faces, the divergence theorem, $v_f^- \leq 0$ for all faces, and the assumption of the proof, we obtain

$$\begin{aligned}
\sum_{f \in \mathcal{F}(\omega)} q_{h,\mathbf{v}} \leq \sum_{f \in \mathcal{F}(\omega)\setminus\mathcal{F}_{in}(\omega)} (v_f^+ + v_f^-)\, p_\omega + \sum_{f \in \mathcal{F}(\omega)\cap\mathcal{F}_{in}(\omega)} v_f^- p_\omega &= p_\omega \sum_{f \in \mathcal{F}(\omega)} v_f \\
&= p_\omega \int_\omega \operatorname{div}\mathbf{v}\, d\omega \leq 0.
\end{aligned}$$

The definition of the diffusion fluxes (2.47), (2.50), (2.52) with non-negative coefficients $T_{\pm,k}$ and assumption that cell p_ω provides a global minimum p_ω immediately imply that

$$\sum_{f \in \mathcal{F}(\omega)} q_h \leq 0.$$

However, the theorem assumption $g \geq 0$ and (2.4) give

$$\sum_{f \in \mathcal{F}(\omega)} (q_h + q_{h,\mathbf{v}}) \geq 0.$$

Therefore,

$$\sum_{f \in \mathcal{F}(\omega)} (q_h + q_{h,\mathbf{v}}) = 0$$

and due to non-positivity of all constituents of the diffusive fluxes in the global minimum we obtain that these constituents are zero and $p_\omega = p_{\omega'}$ for all neighboring cells ω' having a non-zero diffusion matrix connection with ω. The assumption of matrix irreducibility implies that $\mathbf{p}$ is a constant vector. Zero diffusion flux for a boundary face f implies $p_\omega = p_f$, which contradicts with our assumption and proves the assertion of the theorem.

2.9 Numerical Features of Monotone FV Methods

This section addresses only monotonicity properties of the FV schemes for diffusion and convection–diffusion problems. Numerical convergence of these schemes on general 2D and 3D meshes was studied in detail in [48, 51, 106–108, 114, 115, 143]: except for pathological cases, the schemes exhibit approximately the second-order convergence for p and the first-order convergence for q.

We start with the diffusion equation (2.1) defined in the unit cube with a cubic hole, $\Omega = (0,1)^3/[0.4, 0.6]^3$. The boundary of Ω consists of two disjoint parts, interior Γ_w and outer Γ_{out}. We take $g = 0$, the anisotropic diffusion tensor

$$\mathbb{K} = \mathbb{R}_{xyz} \begin{pmatrix} 300 & 0 & 0 \\ 0 & 15 & 0 \\ 0 & 0 & 1 \end{pmatrix} \mathbb{R}_{xyz}^T,$$

where $\mathbb{R}_{xyz} = \mathbb{R}_z(-\pi/6)\mathbb{R}_y(-\pi/4)\mathbb{R}_x(-\pi/3)$, $\mathbb{R}_x$, $\mathbb{R}_y$, $\mathbb{R}_z$ are the rotation matrices for x-, y-, and z-axes, respectively. Also, we impose Dirichlet data setting $\alpha = 1$, $\beta = 0$, and $\gamma = \gamma_*$ where $*$ denotes w or out for Γ_w and Γ_{out}, respectively. The condition on Γ_w imitates the presence of a perforated well. For the first test, we set $\gamma_w = 2$, $\gamma_{out} = 0$. According to the maximum principle for elliptic PDEs, the exact solution should be between 0 and 2. Figure 2.2 demonstrates the solution obtained by different schemes (2.8), (2.14), O-scheme [17] on the cubic grid with $h = 1/40$. We observe that the linear two-point scheme (2.8) produces the solution between 0 and

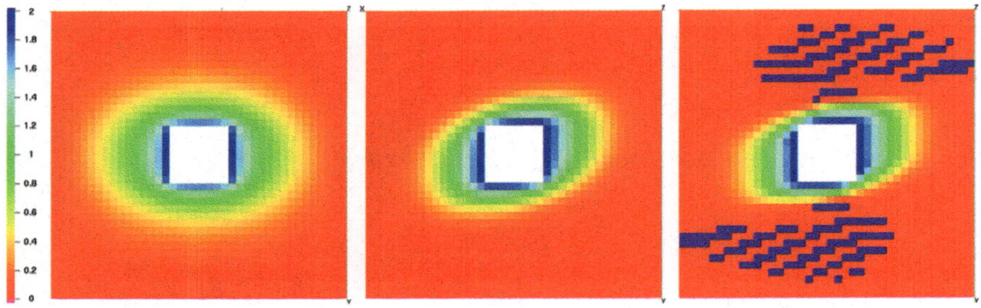

Fig. 2.2 Cutplane of the solution calculated on cubic grid ($h = 1/40$) with the linear TPFA, nonlinear TPFA, and linear MPFA (O-scheme) FV methods for the problem with the Dirichlet boundary conditions. Violet color shows solution less than -10^{-5}. Picture from [113]

Table 2.2 Minimum and maximum of the discrete solution on cubic mesh with $h = 1/40$ for the Dirichlet problem

Scheme	$\gamma_{out} = 0, \gamma_w = 2$		$\gamma_{out} = 10, \gamma_w = 12$	
	p_{min}	p_{max}	p_{min}	p_{max}
lin. TPFA	$1.3 \cdot 10^{-5}$	1.889	10.00	11.889
nonl. TPFA	$1.6 \cdot 10^{-10}$	1.948	9.972	11.940
MPFA [17]	$-5.5 \cdot 10^{-2}$	2.087	9.945	12.087
nonl. MPFA	$1.2 \cdot 10^{-9}$	1.993	10.00	11.993

2 although it is not correct since it is not stretched in the direction of anisotropy. The other linear scheme (O-scheme MPFA) provides approximation although violates the discrete maximum principle. The nonlinear scheme provides both approximation and the discrete maximum principle.

Actually, the nonlinear scheme (2.14) guarantees only solution positivity. Belonging of the solution to the interval (0, 2) is just fortunate. Indeed, if we modify the boundary conditions $\gamma_w = 12$, $\gamma_{out} = 10$, we observe that $p_{\min} < 10$ for the scheme (2.14), see Table 2.2. In this table, we also included extrema of the discrete solution computed by the nonlinear scheme (2.47), (2.50), (2.52) which is proven to satisfy the discrete maximum principle.

The second example imitates the presence of production and injection wells. We consider the diffusion equation (2.1) with $g = 0$ defined in the unit cube with two holes with centrelines $x = 7/22$, $y = 0.5$ and $x = 15/22$, $y = 0.5$, respectively, each with cross section $1/11 \times 1/11$ and height 1. The boundary of Ω consists of three disjoint parts, two interior $\Gamma_{w,1}$, $\Gamma_{w,2}$, and one outer Γ_{out}. On the interior parts, we set $\alpha = 1$, $\beta = 0$, $\gamma = 0$ on $\Gamma_{w,1}$, $\gamma = 1$ on $\Gamma_{w,2}$. On the outer part, we set $\alpha = 0$, $\beta = 1$, $\gamma = 0$ which corresponds to the no-flow boundary condition. The coarsest cubic grid recovering the boundary has mesh size $h = 1/11$, see Fig. 2.3.

The anisotropic diffusion tensor is

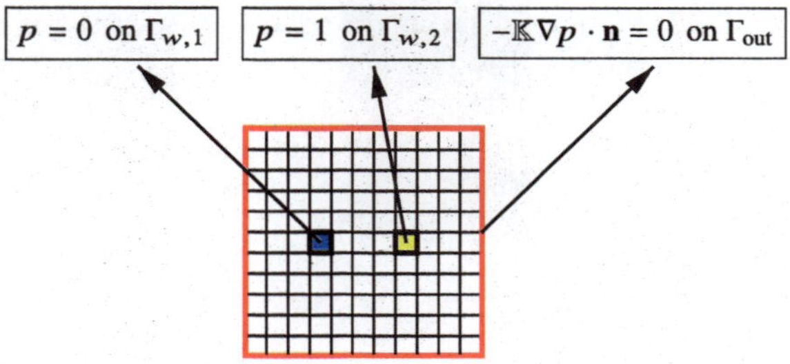

Fig. 2.3 Cutplane of the coarsest cubic grid ($h = 1/11$) for the two-well test case

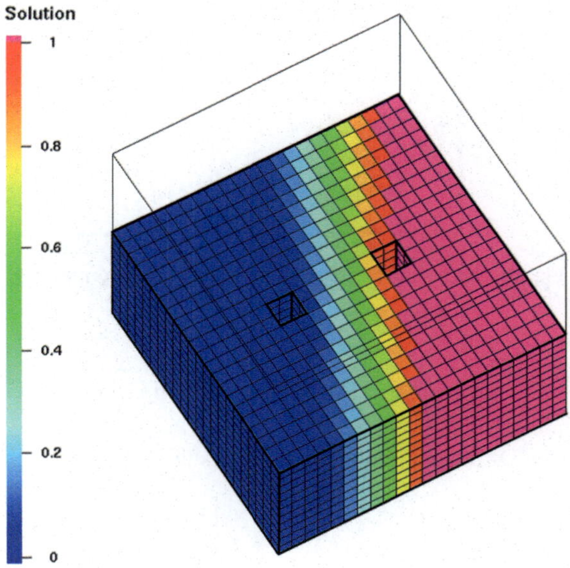

Fig. 2.4 The solution on cubic grid ($h = 1/22$) produced by the nonlinear TPFA scheme (2.14) for the two-well test case

$$\mathbb{K} = \mathbb{R}_z(-67.5°) \begin{pmatrix} 1 & 0 & 0 \\ 0 & 10^{-3} & 0 \\ 0 & 0 & 1 \end{pmatrix} \mathbb{R}_z(67.5°).$$

According to the discrete maximum principle, the solution belongs to the interval [0, 1]. However, the nonlinear scheme with TPFA (2.14) provides only positivity of the solution as shown in Fig. 2.4, whereas the nonlinear scheme with MPFA (2.47), (2.50), (2.52) produces the discrete solution from [0, 1].

Table 2.3 demonstrates that refining the grid reduces the overshoots in the solution of (2.1) generated by the scheme (2.14) since the scheme is consistent.

For the third example, we consider the 2D singularly perturbed convection–diffusion equation (2.62) in the unit square with discontinuous Dirichlet boundary data producing an internal shock in the solution, in addition to exponential boundary layers (Fig. 2.5). Following [85], we set

Table 2.3 Maximum of the discrete solution on the cubic meshes with $h = 1/11, 1/22, 1/44, 1/88$ for the two-well problem

h	$1/11$	$1/22$	$1/44$	$1/88$
nonl.TPFA	2.163	1.765	1.136	1.024
nonl.MPFA	0.992	0.999	0.999	0.999

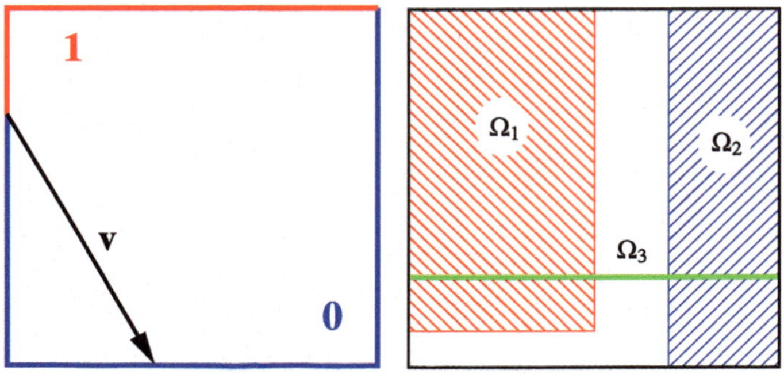

Fig. 2.5 Left: boundary conditions and velocity direction, right: location of Ω_1, Ω_2 and Ω_3 domains

$$\mathbf{v} = \left(\cos\frac{\pi}{3}, -\sin\frac{\pi}{3}\right), \qquad \mathbb{K} = 10^{-8}\mathbb{I}.$$

The Dirichlet boundary conditions are

$$p(x, y) = \begin{cases} 0 & \text{if } x = 1 \text{ or } y \le 0.7, \\ 1 & \text{otherwise,} \end{cases}$$

and therefore, the exact solution has an internal layer along the velocity streamline passing through point $(0,\ 0.7)$ and boundary layers next to two boundary lines $y = 0$ and $x = 1$. The differential solution belongs to the interval $[0, 1]$.

Unstructured polygonal and triangular meshes have the effective mesh parameter $h = 1/64$, so the mesh Péclet number is $Pe = 781,250$. The coarser analogs with $h = 1/32$ are shown in Fig. 2.6.

For the sake of comparison of numerical solutions, the authors of [85] have proposed several metrics of numerical oscillations and smearing effects in subdomains $\Omega_1 = \{(x, y) \in \Omega : x \le 0.5, y \ge 0.1\}$, $\Omega_2 = \{(x, y) \in \Omega : x \ge 0.7\}$, and a cell strip in the vicinity of the line $y = 0.25$, $\Omega_3 = \{\omega \in \Omega^h : \mathbf{x}_\omega = (x_\omega, y_\omega), |y_\omega - 0.25| \le |\omega|^{1/2}\}$. Undershoots and overshoots in Ω_1 are measured by

$$osc_{\text{int}} \equiv \Big(\sum_{\mathbf{x}_\omega \in \Omega_1} (\min\{0,\ p_\omega\})^2 + (\max\{0,\ p_\omega - 1\})^2 \Big)^{1/2}, \tag{2.111}$$

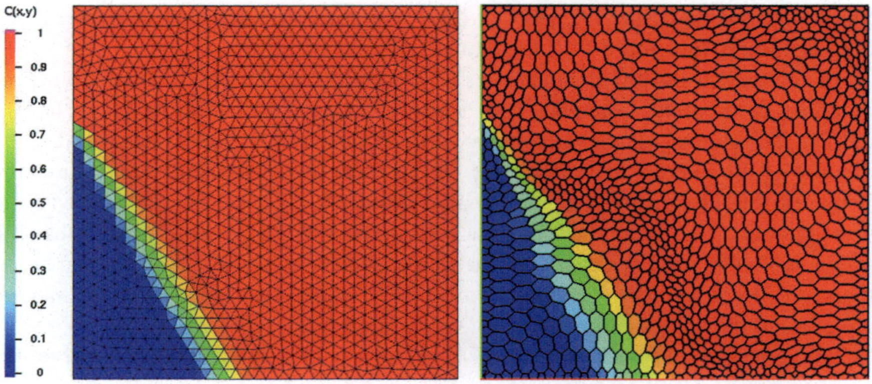

Fig. 2.6 Unstructured meshes and the numerical solution of (2.62). Effective mesh size $h = 1/32$

Table 2.4 Measures of oscillations and diffusive smearing of the discrete solutions to the (2.62) on meshes with effective mesh size $h = 1/64$

Name	osc_{int}	osc_{exp}	$smear_{\text{int}}$	$smear_{\text{exp}}$
Polygonal mesh				
nonl.TPFA	7.0e-8	1.8e-13	1.1e-1	8.4e-5
nonl.MPFA	1.9e-15	3.0e-15	1.1e-1	7.0e-5
Triangular mesh				
nonl.TPFA	3.5e-6	5.0e-7	5.9e-2	2.2e-5
SUPG [41]	5.9e-1	1.5e-0	5.5e-2	4.1e-1
MH85 [84]	4.9e-15	1.8e-14	9.7e-2	5.3e-2

oscillations near the boundary layer in Ω_2 are measured by

$$osc_{\text{exp}} \equiv \Big(\sum_{\mathbf{x}_\omega \in \Omega_2} (\max\{0,\, p_\omega - 1\})^2 \Big)^{1/2}, \tag{2.112}$$

the thickness of the boundary layer and the internal shock are measured by

$$smear_{\text{exp}} \equiv \Big(\sum_{\mathbf{x}_\omega \in \Omega_2} (\min\{0,\, p_\omega - 1\})^2 \Big)^{1/2}, \tag{2.113}$$

$$smear_{\text{int}} \equiv x_2 - x_1, \tag{2.114}$$

where

$$x_1 = \min_{\mathbf{x}_\omega \in \Omega_3,\, p_\omega \ge 0.1} x_T \qquad \text{and} \qquad x_2 = \max_{\mathbf{x}_\omega \in \Omega_3,\, p_\omega \le 0.9} x_T .$$

Small values of measures (2.111)–(2.114) of a numerical solution characterize its non-oscillatory and non-diffusive features.

In Table 2.4, we show the measures (2.111)–(2.114) of numerical solutions provided by the nonlinear schemes with TPFA (2.14) and with MPFA (2.47), (2.50), (2.52). The relatively large value of $smear_{\text{int}}$ is caused by the very coarse grid in the vicinity of the internal layer. We note that the best finite element results presented in review [85] demonstrate slightly smaller smear-out on triangular meshes with the same effective mesh size: the triangular cells in the internal layer are smaller than the polygonal cells, see Fig. 2.6. Other metrics of the FV numerical solution are better than for FE competitors. The metrics of the FV solutions on the triangular meshes are close to the values obtained on the polygonal mesh.

We conclude that the nonlinear FV schemes with two-point (2.14) and multi-point (2.47), (2.50), (2.52) flux approximations demonstrate their competitiveness with the best FE methods regarding minimal oscillations and smearing out of internal and boundary layers.

The use of the flux discretization (2.37) instead of (2.14) for the diffusion problem produces results similar to those reported in Tables 2.2 and 2.3 [143].

The flux vector discretization (2.100), (2.105) for the Navier–Stokes equations demonstrates oscillation-free features as well as small numerical viscosity on several benchmarks in spite of lesser than the second order of approximation, rf. to Chap. 5 and [142].

Chapter 3
Application of MFV in Reservoir Simulation

The chapter is devoted to application of the monotone FM methods in reservoir simulation. We will consider single-phase and multi-phase black-oil models, flows in fractured media, and well-driven flows. The black-oil model is the set of PDEs that describe subsurface flow during the oil and gas recovery from natural subsurface reservoirs. The numerical modeling is the primary decision-making tool for well drilling and management. Geological surveys, core analysis, ultrasound reconnaissance, and laboratory tests are mandatory steps preceding the reservoir simulation. Reservoir simulation implies multiple numerical tests with various scenarios to coin out the best strategy for management of a particular reservoir. Multiple runs of the simulator require its computational efficiency and physically correct results. Monotone FV schemes facilitate achieving these goals.

3.1 Subsurface Flow Models

Here, we present the most common flow models used in the reservoir simulation: single-phase, two-phase, and three-phase black-oil models [26, 47].

3.1.1 Single-Phase Flow

The simplest of the groundwater flow models is the single-phase flow model. Let $\Omega \in \mathbb{R}^3$ be the computational domain for the reservoir, and then the equation for the unknown pressure p is [47]

$$\varphi \rho C_R \frac{\partial p}{\partial t} - \text{div}\left(\frac{\rho}{\mu}\mathbb{K}(\nabla p - \rho g \nabla z)\right) = q, \tag{3.1}$$

where $\mathbb{K}$ is the permeability tensor, φ is the porosity, C_R is the rock compressibility, ρ is the density, μ is the viscosity, g is the gravity constant, z is the depth, and q is the well term.

Y. Vassilevski et al., *Parallel Finite Volume Computation on General Meshes*,
https://doi.org/10.1007/978-3-030-47232-0_3

Equation (3.1) is nonlinear as both density and viscosity depend on the unknown pressure: $\rho = \rho(p)$, $\mu = \mu(p)$.

3.1.2 *Two-Phase Flow*

Next, we introduce basic equations for the two-phase flow model, which is often used to describe an oil–water system. The phase, which wets the medium more than the other, is called wetting phase and is indicated by subscript w. The other phase is the non-wetting phase and indicated by o.

The basic equations for the two-phase flow are the following:

1. Mass conservation for each phase:

$$\frac{\partial \rho_\alpha \varphi S_\alpha}{\partial t} + \mathrm{div}(\rho_\alpha \mathbf{u}_\alpha) = q_\alpha, \quad \alpha = w, o; \tag{3.2}$$

2. Darcy's law:

$$\mathbf{u}_\alpha = -\lambda_\alpha \mathbb{K} \left(\nabla p_\alpha - \rho_\alpha g \nabla z\right), \quad \alpha = w, o; \tag{3.3}$$

3. Two fluids fill the voids:

$$S_w + S_o = 1; \tag{3.4}$$

4. Pressure difference between phases is given by capillary pressure $p_c = p_c(S_w)$:

$$p_o - p_w = p_c. \tag{3.5}$$

Here $\mathbb{K}$ is the absolute permeability tensor, φ is the porosity, g is the gravity constant, z is the depth; for phase α: p_α is the unknown pressure, S_α is the unknown saturation, q_α is the source/sink well term, $\mathbf{u}_\alpha$ is the unknown Darcy's velocity, ρ_α is the unknown density, $B_\alpha = \rho_{\alpha,0}/\rho_\alpha$ is the formation volume factor, $\lambda_\alpha = k_{r\alpha}/(\mu_\alpha B_\alpha)$ is the mobility, μ_α is the viscosity, and $k_{r\alpha}$ is the relative phase permeability.

We choose oil pressure p_o and water saturation S_w as primary unknowns $(p, S) \equiv (p_o, S_w)$. In the sequel, we also take into account the following constitutive relations:

$$k_{r\alpha} = k_{r\alpha}(S_w), \quad \mu_\alpha = \mu_\alpha(p_o), \quad B_\alpha = B_\alpha(p_o), \quad \varphi = \varphi_0\left(1 + C_R(p_o - p_o^0)\right),$$

where C_R is the rock compressibility constant.

3.1.3 *Three-Phase Flow*

The three-phase model describes the flow in the water–oil–gas system. The water is not mixed with two other phases, but the gas can dissolve into the oil phase.

The saturation of the porous rock by phase α is represented by $S_\alpha \in [0, 1]$, where $\alpha = w, o, g$ for water, oil, and gas, respectively. We assume that the mixture fills all the voids:

$$S_w + S_o + S_g = 1. \tag{3.6}$$

The flow of the three-phase mixture through a heterogeneous porous rock is guided by the system:

$$\begin{aligned} \frac{\partial \rho_w \varphi S_w}{\partial t} &+ \text{div}\ (\rho_w \mathbf{u}_w) = q_w, \\ \frac{\partial \rho_o \varphi S_o}{\partial t} &+ \text{div}\ (\rho_o \mathbf{u}_o) = q_o, \\ \frac{\partial \left(\rho_{go} \varphi S_o + \rho_g \varphi S_g\right)}{\partial t} &+ \text{div}\ \left(\rho_{go} \mathbf{u}_o + \rho_g \mathbf{u}_g\right) = q_g, \end{aligned} \tag{3.7}$$

with Darcy fluxes:

$$\begin{aligned} \mathbf{u}_w &= -\lambda_w \mathbb{K} \left(\nabla p_w - \rho_w g \nabla z\right), \\ \mathbf{u}_o &= -\lambda_o \mathbb{K} \left(\nabla p_o - (\rho_o + \rho_g) g \nabla z\right), \\ \mathbf{u}_g &= -\lambda_g \mathbb{K} \left(\nabla p_g - \rho_g g \nabla z\right). \end{aligned} \tag{3.8}$$

Here, φ is the rock porosity, $\rho_\alpha = \rho_{\alpha,0}/B_\alpha$ is the phase density, $\rho_{\alpha,0}$ is the phase density at surface conditions, B_α is the phase formation volume factor, $\rho_{go} = \rho_{g,0} Rs/B_o$ is the density of gas dissolved in oil, Rs is the solubility of gas in the oil at a given bubble point pressure p_b, $\mathbb{K}$ is the absolute permeability tensor, $\lambda_\alpha = k_{r\alpha}/(\mu_\alpha B_\alpha)$ is the phase mobility, $k_{r\alpha}$ is the phase relative permeability, μ_α is the phase viscosity, g is the gravity constant, z is the depth, and q_α is the source or sink for phase $\alpha = w, o, g$.

There are three unknowns in the model, associated with three equations (3.7)–(3.8):

- the oil pressure $p = p_o$,
- the water saturation S_w, and
- an unknown Y that represents either the gas saturation S_g or the bubble point pressure p_b.

The meaning of the unknown Y depends on the state of the mixture, $Y = S_g$, if the gas phase is present, otherwise $Y = p_b$, if gas is fully dissolved in the oil phase. When the gas phase is present, we assume for the bubble point pressure $p_b = p$. The water and gas pressures are connected to the oil pressure p through the water–oil and gas–oil capillary pressures $Pc_{ow}(S_w)$ and $Pc_{go}(S_g)$, functions of water and gas saturations, respectively:

$$p_w = p - Pc_{ow}(S_w), \quad p_g = p + Pc_{go}(S_g). \tag{3.9}$$

We introduce the following constitutive relations for (3.7)–(3.8).

The rock porosity $\varphi = \varphi(p)$ depends on the pressure:

$$\varphi(p) = \varphi_0 \left(1 + C_R \left(p - p_0\right)\right), \tag{3.10}$$

where C_R is the rock compressibility constant, and $\varphi_0 = \varphi(p_0)$ is the rock porosity at pressure p_0.

The densities are expressed through the formation volume factor which is the function of the pressure:

- $\rho_w(p) = \rho_{w,0}/B_w(p)$ for the density of water,
- $\rho_o(p_b) = \rho_{o,0}/B_o(p_b)$ for the density of oil that depends on the bubble point pressure,
- $\rho_g(p) = \rho_{g,0}/B_g(p)$ for the density of gas,
- $\rho_{go} = \rho_{g,0}Rs(p_b)/B_o(p_b)$ for the density of gas dissolved in oil,
- $Rs(p_b)$ describes the solubility of gas in oil at a given bubble point pressure p_b.

For the relative permeabilities, we have

- $k_{rw}(S_w)$ for water,
- $k_{rg}(S_g)$ for gas,
- $k_{ro}(S_w, S_g) = \left(k_{rog}(S_g) + k_{rg}(S_g)\right)\left(k_{row}(S_w) + k_{rw}(S_w)\right)/k_{rowc} - (k_{rw}(S_w) - k_{rg}(S_g))$ for oil, according to second Stone's model,
- $k_{row}(S_w)$ for oil at $S_g = S_{gc}$, the critical gas saturation,
- $k_{rog}(S_g)$ for oil at $S_w = S_{wc}$, the connate water saturation, and
- k_{rowc} for oil at the connate water saturation S_{wc} and the critical gas saturation S_{gc}.

Further, the phase mobilities that account for change in the relative permeability and the viscosity are

- $\lambda_w(S_w, p) = k_{rw}(S_w)/\mu_w(p)$ for water,
- $\lambda_g(S_g, p) = k_{rg}(S_g)/\mu_g(p)$ for gas, and
- $\lambda_o(S_w, S_g, p_b) = k_{ro}(S_w, S_g)/\mu_o(p_b)$ for oil.

Therefore, the Darcy velocities for water $\mathbf{u}_w$, oil $\mathbf{u}_o$, free gas $\mathbf{u}_g$, and dissolved gas $\mathbf{u}_{go}$ are

$$\begin{aligned} \mathbf{u}_w &= \lambda_w \mathbf{v}_w := -\lambda_w \mathbb{K}\left(\nabla p - \nabla Pc_{ow} - \rho_w g \nabla z\right), \\ \mathbf{u}_o &= \lambda_o \mathbf{v}_o := -\lambda_o \mathbb{K}\left(\nabla p - (\rho_o + \rho_g) g \nabla z\right), \\ \mathbf{u}_g &= \lambda_g \mathbf{v}_g := -\lambda_g \mathbb{K}\left(\nabla p + \nabla Pc_{go} - \rho_g g \nabla z\right), \end{aligned} \tag{3.11}$$

where $\mathbf{v}_\alpha, \alpha = w, o, g$ are velocities without the mobility.

The water volume factor and viscosity properties are defined by

$$\begin{aligned} B_w(p) &= \frac{B_{w,0}}{1 + C_B(p - p_0) + \frac{1}{2}C_B^2(p - p_0)^2}, \\ \mu_w(p) &= \frac{\mu_{w,0}}{1 + C_\mu(p - p_0) + \frac{1}{2}C_\mu^2(p - p_0)^2}, \end{aligned} \tag{3.12}$$

where $B_{w,0}$ is the default water formation volume factor and $\mu_{w,0}$ is the viscosity, C_B is the water compressibility, and C_μ is the viscosibility. The oil and gas volume factors, B_o and B_g, viscosities μ_o and μ_g, capillary pressures Pc_{wo}, and Pc_{go} are set by key–value tables, where the key is the pressure. The relative permeability for water k_{rw}, gas k_{rg}, and oil with respect to connate water k_{row} and critical gas k_{rog} are also provided by key–value tables, where the key is the saturation.

3.1.4 Well Model and Boundary Conditions

On the reservoir boundaries, the no-flow (homogeneous Neumann) boundary condition is imposed. Therefore, inflow and outflow are performed through wells.

Wells are incorporated through the well terms in (3.1), (3.2), or (3.7). Each well is assumed to be connected to the center of a cell. The formula for the well term was suggested by Peaceman [123]. For a cell ω with center $\mathbf{x}_\omega$ connected to the well, we have

$$q_\alpha = \frac{\rho_\alpha k_{r\alpha}}{\mu_\alpha} WI\Big(p_{bh} - p - \rho_\alpha(z_{bh} - z)\Big)\delta(\mathbf{x} - \mathbf{x}_\omega), \tag{3.13}$$

where p_{bh} is the bottom hole pressure, z_{bh} is the depth of the bottom hole, δ is the Dirac delta function, and WI is the well index, which does not depend on the properties of fluids, but depends on properties of the medium:

$$WI = \frac{2\pi h_z \sqrt{K_x K_y}}{\log\,(r_0/r_w) + s}, \quad r_0 = 0.28 \frac{\sqrt{\sqrt{\frac{K_y}{K_x}} h_x^2 + \sqrt{\frac{K_x}{K_y}} h_y^2}}{\left(\frac{K_y}{K_x}\right)^{\frac{1}{4}} + \left(\frac{K_x}{K_y}\right)^{\frac{1}{4}}}. \tag{3.14}$$

Here, for a general cell ω and tensor permeability $\mathbb{K}$ we use eigendecomposition $\mathbb{K} = R\ \mathrm{diag}\ \left(K_x, K_y, K_z\right) R^T$, define h_x, h_y, h_z as dimensions of the box aligned with the axes rotated by R and bounding cell ω, r_0 is the equivalent radius, r_w is the well radius, and s is the skin factor. A positive source term leads to fluid injection or a negative source term leads to fluid production.

We also assume that there is no capillary pressure in wells, so all fluxes depend on the same (oil) pressure.

As a possible alternative to the well index approach, one may also consider a near-well correction method presented in Sect. 3.5.

3.2 Time-Stepping and Nonlinear Systems

There are several methods for time discretization of the two-phase and three-phase flow equations, the most popular are IMPES (Implicit Pressure—Explicit Saturation), sequential (Implicit Pressure—Implicit Saturation), and fully implicit schemes. Two of these methods will be discussed below.

3.2.1 IMPES Scheme for Two-Phase Flow

In this section, we present the IMPES time-stepping for the two-phase flow model. The method is simple to implement, but the explicit step for the saturation restricts the size of the time step.

Let the total velocity be $\mathbf{u} = \mathbf{u}_w + \mathbf{u}_o$. If the rock porosity and liquid densities are fixed during the time step, then $\frac{\partial \rho_w \varphi S_w}{\partial t} + \frac{\partial \rho_o \varphi S_o}{\partial t} = 0$. Dividing equations (3.2) by ρ_α and adding them together results in

$$\operatorname{div}(\mathbf{u}) = q_w/\rho_w + q_o/\rho_o. \tag{3.15}$$

Applying (3.5) to (3.3) gives

$$\mathbf{u} = -\mathbb{K}\big(\lambda \nabla p - \lambda_w \nabla p_c - (\lambda_w \rho_w + \lambda_o \rho_o) g \nabla z\big), \tag{3.16}$$

where $\lambda = \lambda_w + \lambda_o$ is the total mobility.

Substituting (3.16) into (3.15) gives the pressure equation

$$-\operatorname{div}(\mathbb{K}\lambda \nabla p) = q_w/\rho_w + q_o/\rho_o - \operatorname{div}\big[\mathbb{K}\big(\lambda_w \nabla p_c + (\lambda_w \rho_w + \lambda_o \rho_o) g \nabla z\big)\big]. \tag{3.17}$$

The phase velocities $\mathbf{u}_w$ and $\mathbf{u}_o$ can be expressed through the total velocity $\mathbf{u}$ by

$$\mathbf{u}_w = \frac{\lambda_w}{\lambda}\big(\mathbf{u} + \mathbb{K}\lambda_o \nabla p_c + \mathbb{K}\lambda_o(\rho_w - \rho_o) g \nabla z\big),$$

$$\mathbf{u}_o = \frac{\lambda_o}{\lambda}\big(\mathbf{u} - \mathbb{K}\lambda_w \nabla p_c + \mathbb{K}\lambda_w(\rho_o - \rho_w) g \nabla z\big).$$

Similarly, (3.5) and (3.16) applied to (3.2) and (3.3) (for $\alpha = w$) yield the saturation equation

$$\operatorname{div}\frac{\lambda_w}{\lambda}\left[\mathbf{u} + \mathbb{K}\lambda_o\left(\frac{dp_c}{dS}\nabla S + (\rho_o - \rho_w) g \nabla z\right)\right] + \varphi \frac{\partial S}{\partial t} = q_w/\rho_w. \tag{3.18}$$

Finally, the IMPES method can be formalized:

1. Solve **implicitly** (3.17) to obtain current pressure p^n from current saturation S^n:

$$
\begin{aligned}
- \operatorname{div}(\mathbb{K}\lambda^n \nabla p^n) &= q_w/\rho_w + q_o/\rho_o \\
&= -\operatorname{div}\big[\mathbb{K}\big(\lambda_w^n \nabla p_c + (\lambda_w^n \rho_w^n + \lambda_o^n \rho_o^n) g \nabla z\big)\big],
\end{aligned}
\tag{3.19}
$$

where $\lambda_\alpha^n = \lambda_\alpha(S^n, p^n)$ and $\rho_\alpha^n = \rho_\alpha(p^n)$.

2. Use (3.16) to find current Darcy's velocity $\mathbf{u}^n$ using current S^n and p^n:

$$
\mathbf{u}^n = -\mathbb{K}\big(\lambda^n \nabla p^n - \lambda_w^n \nabla p_c - (\lambda_w^n \rho_w^n + \lambda_o^n \rho_o^n) g \nabla z\big). \tag{3.20}
$$

3. Solve **explicitly** (3.18) to get the next time step saturation S^{n+1} using current S^n, p^n, and $\mathbf{u}^n$:

$$
\begin{aligned}
&\frac{1}{\Delta t^{n+1}} \left[\left(\frac{\varphi S}{B_w} \right)^{n+1} - \left(\frac{\varphi S}{B_w} \right)^{n} \right] \\
&= q_w/\rho_w - \operatorname{div} \frac{\lambda_w^n}{\lambda^n} \left[\mathbb{K}\lambda_o^n \left(\frac{dp_c}{dS} \nabla S + (\rho_o^n - \rho_w^n) g \nabla z \right) + \mathbf{u}^n \right].
\end{aligned}
\tag{3.21}
$$

Note that Eq. (3.19) is the steady diffusion equation with diffusion tensor $\lambda^n \mathbb{K}$. Discretization of the diffusion fluxes in (3.19), as well as computation of the right-hand sides in (3.20) and (3.21) is based on the nonlinear TPFA or MPFA from Chap. 2. The use of the Picard method for the solution of (3.19) with lagging coefficients from previous nonlinear iteration allows us to take all coefficients in (3.19) implicitly: λ^n, λ_w^n, λ_o^n, ρ_w^n, and ρ_o^n.

3.2.2 *Fully Implicit Scheme for Three-Phase Flow*

Another time discretization technique widely used in reservoir simulation is the fully implicit backward Euler method. The discretization for accumulation terms in (3.7) in cell ω takes the form:

$$
\begin{aligned}
R_{w,\omega}^a &:= \int_\omega \frac{\partial \rho_w \varphi S_w}{\partial t} d\omega \approx \frac{\rho_{w,0}|\omega|}{\Delta t} \left(\frac{\varphi(p^{n+1}) S_w^{n+1}}{B_w(p^{n+1})} - \frac{\varphi(p^n) S_w^n}{B_w(p^n)} \right), \\
R_{o,\omega}^a &:= \int_\omega \frac{\partial \rho_o \varphi S_o}{\partial t} d\omega \approx \frac{\rho_{o,0}|\omega|}{\Delta t} \left(\frac{\varphi(p^{n+1}) S_o^{n+1}}{B_o(p_b^{n+1})} - \frac{\varphi(p^n) S_o^n}{B_o(p_b^n)} \right), \\
R_{g,\omega}^a &:= \int_\omega \frac{\partial \left(\rho_{go} \varphi S_o + \rho_g \varphi S_g\right)}{\partial t} d\omega \approx \frac{\rho_{g,0}|\omega|}{\Delta t} \left(\frac{\varphi(p^{n+1}) Rs(p_b^{n+1}) S_o^{n+1}}{B_o(p_b^{n+1})} - \frac{\varphi(p^n) Rs(p_b^n) S_o^n}{B_o(p_b^n)} \right) \\
&\quad + \frac{\rho_{g,0}|\omega|}{\Delta t} \left(\frac{\varphi(p^{n+1}) S_g^{n+1}}{B_g(p^{n+1})} - \frac{\varphi(p^n) S_g^n}{B_g(p^n)} \right).
\end{aligned}
\tag{3.22}
$$

The transport terms in (3.7) are discretized with the finite volume method and are expressed by

$$
\begin{aligned}
R^t_{w,\omega} &:= \int_\omega \text{div } (\rho_w \mathbf{u}_w)\, \mathrm{d}\omega \approx \sum_{f \in \mathcal{F}(\omega)} \rho_w(p^{n+1}) \lambda_w(S_w^{n+1}, p^{n+1}) \mathbf{n}^T \mathbf{v}_w |f|, \\
R^t_{o,\omega} &:= \int_\omega \text{div } (\rho_o \mathbf{u}_o)\, \mathrm{d}\omega \approx \sum_{f \in \mathcal{F}(\omega)} \rho_o(p_b^{n+1}) \lambda_o(S_w^{n+1}, S_g^{n+1}, p_b^{n+1}) \mathbf{n}^T \mathbf{v}_o |f|, \\
R^t_{g,\omega} &:= \int_\omega \text{div } \left(\rho_g \mathbf{u}_g + \rho_{go} \mathbf{u}_o\right) \mathrm{d}\omega \approx \sum_{f \in \mathcal{F}(\omega)} \rho_g(p^{n+1}) \lambda_g(S_g^{n+1}, p^{n+1}) \mathbf{n}^T \mathbf{v}_g |f| \\
&\qquad + \sum_{f \in \mathcal{F}(\omega)} \rho_{go}(p_b^{n+1}) \lambda_o(S_w^{n+1}, S_g^{n+1}, p^{n+1}) \mathbf{n}^T \mathbf{v}_{go} |f|.
\end{aligned}
\tag{3.23}
$$

We discretize each of the fluxes

$$
\begin{aligned}
q^p &= -\mathbf{n}^T \mathbb{K} \nabla p, \\
q^z &= \mathbf{n}^T \mathbb{K} \nabla z, \\
q^{ow} &= \mathbf{n}^T \mathbb{K} \nabla Pc_{ow}, \\
q^{go} &= -\mathbf{n}^T \mathbb{K} \nabla Pc_{go}
\end{aligned}
\tag{3.24}
$$

using the nonlinear two-point or multi-point flux approximation from Chap. 2 on the next time level, and get the discrete fluxes q_h^p, q_h^z, q_h^{ow}, q_h^{go} (q_h^z is calculated prior the simulation). Then $\mathbf{n}^T \mathbf{v}_\alpha$ are approximated as follows:

$$
\begin{aligned}
\mathbf{n}^T \mathbf{v}_w &\approx q_h^p + q_h^{ow} + \rho_w\, q_h^z, \\
\mathbf{n}^T \mathbf{v}_o &\approx q_h^p + (\rho_o + \rho_g)\, q_h^z, \\
\mathbf{n}^T \mathbf{v}_g &\approx q_h^p + q_h^{go} + \rho_g\, q_h^z.
\end{aligned}
\tag{3.25}
$$

We use the single-point upstream method to discretize the advection in (3.23). Let $f = \omega_+ \cap \omega_-$ be a face with normal $\mathbf{n}$ directed from ω_+ into ω_-, velocity $v = \mathbf{n}^T \mathbf{v}$, and $f(C)$ be advected function with vector of arguments C that has values $C_+ = \left(p_+, p_{b+}, S_{w+}, S_{g+}\right)$ at center of cell ω_+ and $C_- = \left(p_-, p_{b-}, S_{w-}, S_{g-}\right)$ at center of cell ω_-, then we define the discrete function upw $(f(C), v)$ as

$$
\text{upw } (f(C), v) = \begin{cases} f(C_+) v, & v \geq 0, \\ f(C_-) v, & v < 0, \end{cases}
\tag{3.26}
$$

The discretization of the transport term (3.23) is

$$
\begin{aligned}
R^t_{w,\omega} &\approx \sum_{f\in\mathcal{F}(\omega)} \begin{pmatrix} \mathrm{upw}\left(\rho_w(p^{n+1})\lambda_w(S_w^{n+1}, p^{n+1}), v_p\right) + \\ \mathrm{upw}\left(\rho_w(p^{n+1})\lambda_w(S_w^{n+1}, S_g^{n+1}, p_b^{n+1}), v_{Pc_{ow}}\right) + \\ \mathrm{upw}\left(\rho_w^2(p^{n+1})\lambda_w(S_w^{n+1}, p^{n+1}), v_z\right) \end{pmatrix} |f|, \\
R^t_{o,\omega} &\approx \sum_{f\in\mathcal{F}(\omega)} \begin{pmatrix} \mathrm{upw}\left(\rho_o(p_b^{n+1})\lambda_o(S_w^{n+1}, S_g^{n+1}, p_b^{n+1}), v_p\right) + \\ \mathrm{upw}\left(\rho_o(p_b^{n+1})\left(\rho_o(p_b^{n+1}) + \rho_g(p^{n+1})\right)\lambda_o(S_w^{n+1}, S_g^{n+1}, p_b^{n+1}), v_z\right) \end{pmatrix} |f|, \\
R^t_{g,\omega} &\approx \sum_{f\in\mathcal{F}(\omega)} \begin{pmatrix} \mathrm{upw}\left(\rho_g(p^{n+1})\lambda_g(S_g^{n+1}, p^{n+1}), v_p\right) + \\ \mathrm{upw}\left(\rho_g(p^{n+1})\lambda_g(S_g^{n+1}, p^{n+1}), v_{Pc_{wo}}\right) + \\ \mathrm{upw}\left(\rho_g^2(p^{n+1})\lambda_g(S_g^{n+1}, p^{n+1}), v_z\right) \end{pmatrix} |f| \\
&+ \sum_{f\in\mathcal{F}(\omega)} \begin{pmatrix} \mathrm{upw}\left(\rho_{go}(p_b^{n+1})\lambda_o(S_w^{n+1}, S_g^{n+1}, p_b^{n+1}), v_p\right) + \\ \mathrm{upw}\left(\rho_{go}(p_b^{n+1})\left(\rho_o(p_b^{n+1}) + \rho_g(p^{n+1})\right)\lambda_o(S_w^{n+1}, S_g^{n+1}, p_b^{n+1}), v_z\right) \end{pmatrix} |f|.
\end{aligned}
\tag{3.27}
$$

For the source term, we apply the single-point upstream method. At cell ω with pressure p, bubble point pressure p_b, water saturation S_w, gas saturation S_g and at passing through ω well, bottom hole pressure p_{bh}, and saturations S_w^{inj}, S_g^{inj} of the injected mixture we denote

$$
v_{ps} = WI\,(p_{bh} - p)\,, \quad v_{zs} = -WI\,(z_{bh} - z_\omega)\,, \quad v_{s\alpha} = v_{ps} + \rho_\alpha v_{zs}, \tag{3.28}
$$

where the well index is given by the Peaceman formula (3.14). If an inverse flow occurs at a producer well, then the injection concentrations are defined based on the last produced saturations. Let us define the single-point upstream function for the well:

$$
\mathrm{upw}_w\,(f(C), v) = \begin{cases} f(C)v, & v \le 0, \\ f(C_w)v, & v > 0, \end{cases} \tag{3.29}
$$

where $C = \left(p, p_b, S_w, S_g\right)$ is the vector of unknowns of cell ω and $C_w = \left(p_{bh}, p_{bh}, S_w^{inj}, S_g^{inj}\right)$ is the vector of unknowns at the well. Then the discretization of the well source–sink term has the following form:

$$
\begin{aligned}
R^{wl}_{w,\omega} &:= \int_\omega q_w \mathrm{d}\omega \approx \mathrm{upw}_w\left(\rho_w\left(p^{n+1}\right)\lambda_w\left(S_w^{n+1}, p^{n+1}\right), v_{ps}\right) \\
&\qquad + \mathrm{upw}_w\left(\rho_w^2\left(p^{n+1}\right)\lambda_w\left(S_w^{n+1}, p^{n+1}\right), v_{zs}\right), \\
R^{wl}_{o,\omega} &:= \int_\omega q_o \mathrm{d}\omega \approx \mathrm{upw}_w\left(\rho_o\left(p_b^{n+1}\right)\lambda_o\left(S_w^{n+1}, S_g^{n+1}, p_b^{n+1}\right), v_{ps}\right) \\
&\qquad + \mathrm{upw}_w\left(\rho_o^2\left(p_b^{n+1}\right)\lambda_o\left(S_w^{n+1}, S_g^{n+1}, p_b^{n+1}\right), v_{zs}\right), \\
R^{wl}_{g,\omega} &:= \int_\omega q_g \mathrm{d}\omega \approx \mathrm{upw}_w\left(\rho_g\left(p^{n+1}\right)\lambda_g\left(S_g^{n+1}, p^{n+1}\right), v_{ps},\right) \\
&\qquad + \mathrm{upw}_w\left(\rho_g^2\left(p^{n+1}\right)\lambda_g\left(S_g^{n+1}, p^{n+1}\right), v_{zs}\right).
\end{aligned}
\tag{3.30}
$$

We assume that the mobility is zero if the respective saturation is zero. The well may also be guided by the target injection or production rate. In this case, the bottom hole pressure of the well becomes an unknown in the system and we set $p_{bh} = p_{bh}^{n+1}$. The additional unknown is guided by the equation that is formed based on difference between the sum of rates over all well perforations and the target well rate.

We define the nonlinear residual to a quantity evaluated at time step $n+1$ inside grid cell ω:

$$R_{\alpha,\omega} = R^{a}_{\alpha,\omega} + R^{t}_{\alpha,\omega} + R^{wl}_{\alpha,\omega}, \quad \alpha = w, o, g, \tag{3.31}$$

and write the fully implicit discretization of (3.7)

$$R_{\alpha,\omega} = 0, \quad \alpha = w, o, g \tag{3.32}$$

for all grid cells ω at every time step.

The resulting nonlinear system is usually solved by the Newton method[1]:

$$\mathcal{J}(x^l)\delta\mathbf{x}^l = -\mathcal{R}(\mathbf{x}^l), \tag{3.33}$$

$$\mathbf{x}^{l+1} = \mathbf{x}^l + \delta\mathbf{x}^l, \tag{3.34}$$

where $\mathbf{x}^l$ is an approximation to $\mathbf{x}^{n+1}$ on the lth Newton step, $\mathbf{x} = (p_o, S_w, Y)^T$ is the vector of primary unknowns in all grid cells, $\mathcal{R}(\mathbf{x}) = (R_w(\mathbf{x}), R_o(\mathbf{x}), R_g(\mathbf{x}))^T$ is the vector of nonlinear residuals in all grid cells, and $\mathcal{J}$ is the Jacobian matrix:

$$\mathcal{J}(x) = \begin{pmatrix} \frac{\partial R_w}{\partial p_o}(\mathbf{x}) & \frac{\partial R_w}{\partial S_w}(\mathbf{x}) & \frac{\partial R_w}{\partial Y}(\mathbf{x}) \\ \frac{\partial R_o}{\partial p_o}(\mathbf{x}) & \frac{\partial R_o}{\partial S_w}(\mathbf{x}) & \frac{\partial R_o}{\partial Y}(\mathbf{x}) \\ \frac{\partial R_g}{\partial p_o}(\mathbf{x}) & \frac{\partial R_g}{\partial S_w}(\mathbf{x}) & \frac{\partial R_g}{\partial Y}(\mathbf{x}) \end{pmatrix}.$$

The Newton method is terminated when the norm of the residual vector drops below $\varepsilon_{\mathrm{nwt}}$.

The sparsity structure of the Jacobian matrix $\mathcal{J}$ is formed by the stencils of variations for fluxes (3.25) with respect to the primary unknowns. In the nonlinear schemes, coefficients in the flux discretization depend on the primary unknowns. Therefore, variations for fluxes (3.25) have larger stencils than (2.14), (2.37), (2.47), rf. to [113].

One can avoid explicit computation of matrix $\mathcal{J}$ entries. Once the residual vector $\mathcal{R}$ is constructed, the Jacobian $\mathcal{J}$ may be obtained using the automatic differentiation procedures of the INMOST platform, rf. Chap. 6.

---

[1] Hereinafter, we use for the Jacobian matrix and residual vector particular notations $\mathcal{J}$ and $\mathcal{R}$

3.3 Simulation of Waterflood

In this section, we present a few numerical results obtained with the nonlinear TPFA for the two-phase flow model. The accuracy and the computational cost of the method are compared with the ones for the conventional linear TPFA or MPFA. For the sake of brevity, we address here the fully implicit scheme only.

In the first two test cases, we consider pseudo-2D problems on $N \times N \times 1$ hexahedral meshes and use the following rock and fluid properties. The relative permeabilities of fluids $k_{r\alpha}$ are shown in Fig. 3.1, left. The capillary pressure p_c dependence on S_w is presented in Fig. 3.1, right. The viscosities μ_α and the volume factors B_α are set by Table 3.1, and the densities are $\rho_{w,0} \approx 4.331 \cdot 10^{-1}$ psi/ft and $\rho_{o,0} \approx 3.898 \cdot 10^{-1}$ psi/ft. The rock is assumed to be incompressible.

The producer and injector wells are incorporated through the bottom hole pressures. For the injector, it is $p_{bh,inj} = 4100$ psia and for the producer $p_{bh,pr} = 3900$ psia. The well indexes are assumed to be fixed $WI = 10 \frac{\text{bbl} \cdot \text{cp}}{\text{day} \cdot \text{psi}}$.

3.3.1 Non-orthogonal Grids

In the first experiment, we take the uniform $32 \times 32 \times 1$ mesh, fix the first two and last two grid lines (this leaves well-connected cells intact), rotate the central lines toward the wells (angle $\alpha = -30°$) or away from them ($\alpha = 30°$), and interpolate the other lines linearly between the central lines and the boundary (see Fig. 3.2). We run simulations of the waterflood with the linear and nonlinear TPFAs on the

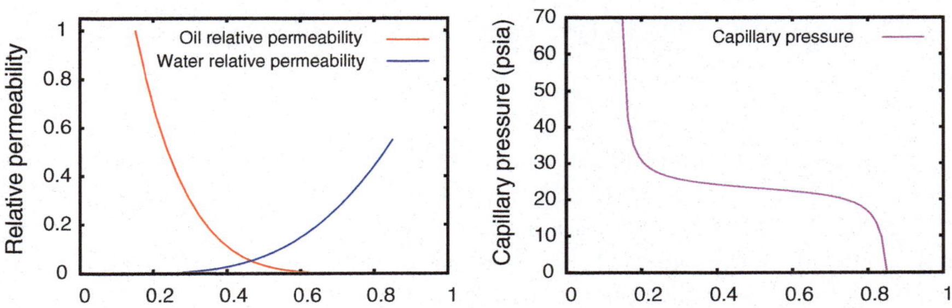

Fig. 3.1 Left: oil and water relative permeabilities. Right: capillary pressure dependence on S_w

Table 3.1 Fluid compressibility properties

p (psia)	B_o ($\frac{\text{bbl}}{\text{STB}}$)	B_w ($\frac{\text{bbl}}{\text{STB}}$)	μ_o (cp)	μ_w (cp)
3900	1.0030	1.01317	90.6	0.515
4000	1.0020	1.01291	96.0	0.518
4100	1.0009	1.01264	101.7	0.521

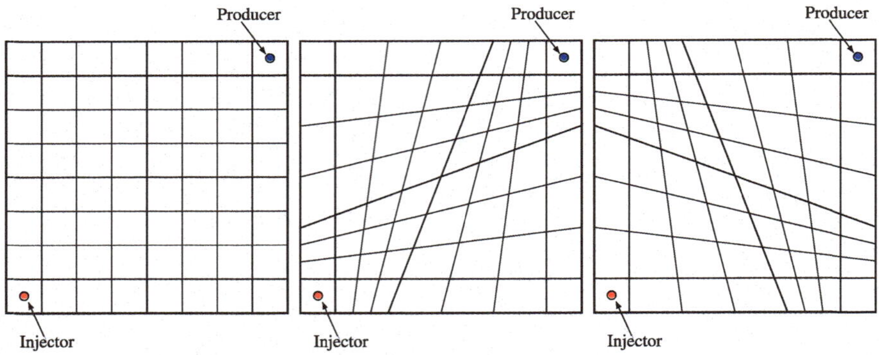

Grid M1 ($\alpha = 0°$) | Grid M2 ($\alpha = -30°$) | Grid M3 ($\alpha = 30°$)

Fig. 3.2 Orthogonal and non-orthogonal sample grids

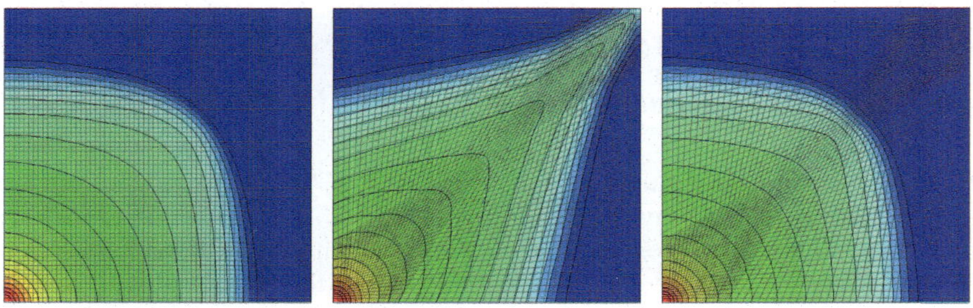

Grid M1 linear=nonlinear | Grid M2 linear | Grid M2 nonlinear

Fig. 3.3 Water saturation for grids M1 and M2 for the linear and nonlinear scheme at $T = 250$ days

modified meshes and compare the results with the ones on the orthogonal mesh. The permeability tensor is chosen to be $\mathbb{K} = \{100, 100, 10\}$.

Figure 3.3 shows the water saturation field on orthogonal grid M1 (Fig. 3.3 left) and on grid M2 for the linear (Fig. 3.3 center) and nonlinear (Fig. 3.3 right) TPFAs. In case of linear discretization, the waterfront accelerates and its form differs from the reference one obtained on the orthogonal grid, while in case of the nonlinear discretization the fronts are almost equal.

Figure 3.4 shows the water saturation field on orthogonal grid M1 (Fig. 3.4 left) and on grid M3 for the linear (Fig. 3.4 center) and nonlinear (Fig. 3.4 right) TPFAs. The nonlinear discretization provides the front which is very close to the front on the orthogonal grid, while the linear TPFA decelerates the waterfront.

The water breakthrough times are also noticeably different for the linear and nonlinear TPFAs on grids M2 and M3 (see Table 3.2). The oil production rates are

Fig. 3.4 Water saturation for grids M1 and M3 for the linear and nonlinear scheme at $T = 250$ days

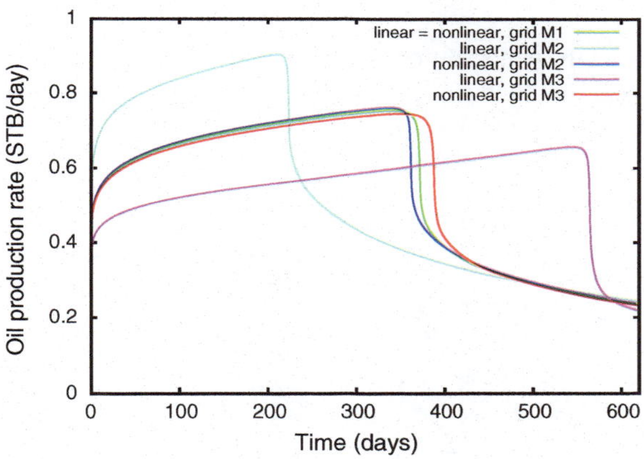

Fig. 3.5 Oil production rates for grids M1, M2, and M3

Table 3.2 Water breakthrough times. Non-orthogonal grid

	Grid M1	Grid M2	Grid M3
Linear TPFA	372.2	224.1	564.2
Nonlinear TPFA	372.2	362.2	388.5

shown in Fig. 3.5. For grid M2, both linear and nonlinear fluxes break earlier than for grid M1, while for grid M3 both linear and nonlinear fluxes alternatively break later than for grid M1. Meanwhile, the breakthrough times for the nonlinear discretization on the modified grids are very close to the reference ones on the orthogonal grid.

3.3.2 Discontinuous Tensor with High Anisotropy

In the next experiment, we use the uniform orthogonal $64 \times 64 \times 1$ grid M1, but with a discontinuous full anisotropic permeability tensor (see Fig. 3.6).

$$\mathbb{K} = \mathbb{R}_z(-\theta)\ \text{diag}(1000, 10, 10)\ \mathbb{R}_z(\theta),$$

where $\text{diag}(k_1, k_2, k_3)$ denotes the diagonal matrix with entries k_1, k_2, k_3, and $\mathbb{R}_\alpha(\theta)$ is the rotation matrix in plane orthogonal to $o\alpha$ with angle θ.

The computational domain is $100\,\text{ft} \times 100\,\text{ft} \times 10\,\text{ft}$, and the rotation angle is the following:

$$\theta = \begin{cases} 0^\circ & \text{if } 50\text{ ft } \le x + y < 100\text{ ft}, \\ 90^\circ & \text{if } 100\text{ ft } \le x + y < 150\text{ ft}, \\ 45^\circ & \text{if } x + y < 50\text{ ft, or } x + y \ge 150\text{ ft}. \end{cases}$$

Figure 3.7 shows the water saturation field at the moment $T = 55$ days, and Fig. 3.8 presents the oil pressure field at the moment $T = 10$ days. The front propagations are completely different even if discretizations differ only near wells, where $\theta = 45^\circ$. Oil and water production rates are shown in Fig. 3.9.

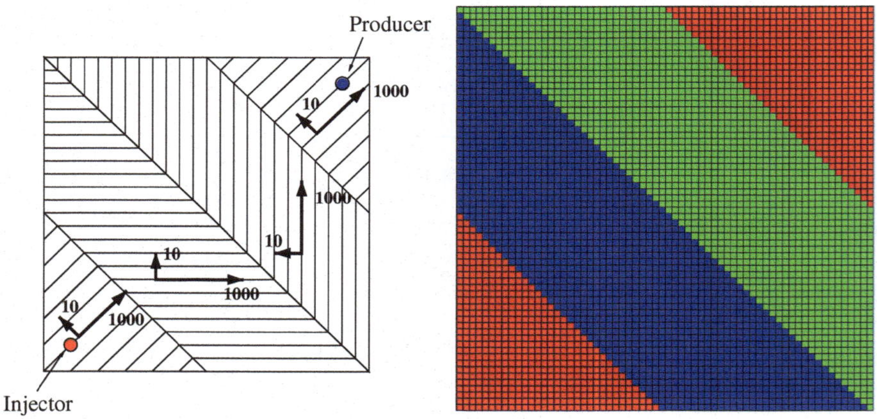

Fig. 3.6 Left: discontinuous anisotropic tensor. Right: sample mesh

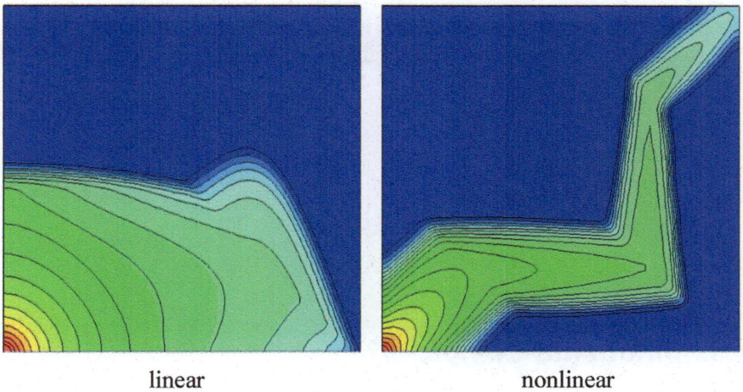

Fig. 3.7 Water saturation at $T = 55$ days. Discontinuous tensor with high anisotropy

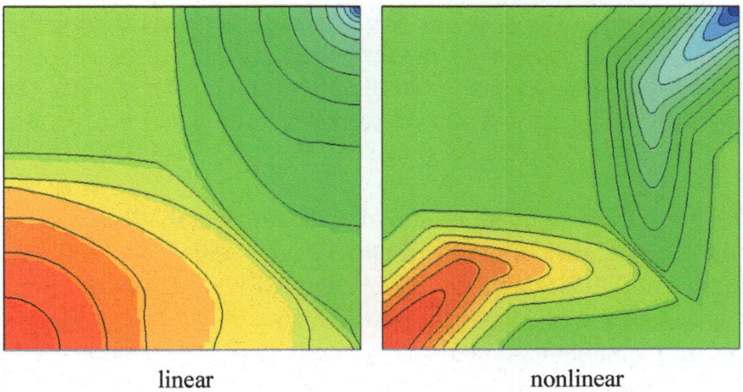

Fig. 3.8 Oil pressure at $T = 10$ days. Discontinuous tensor with high anisotropy

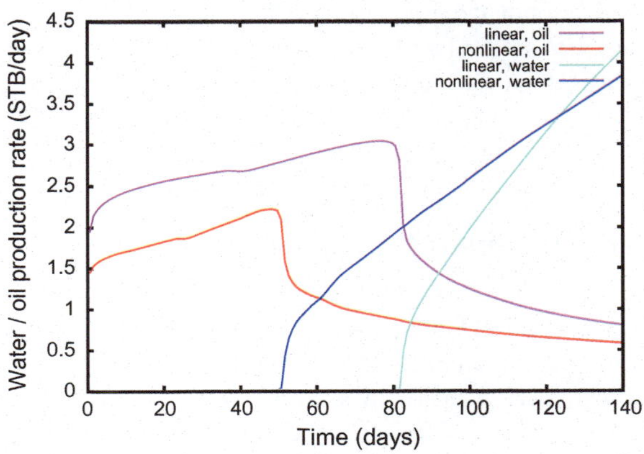

Fig. 3.9 Oil/water production rates. Discontinuous tensor with high anisotropy

Table 3.3 Total number of non-zero elements (nz) in the Jacobian matrix, CPU time (sec), total number of nonlinear (#nit.) and linear (#lit.) iterations for simulations with linear TPFA, nonlinear TPFA, and MPFA-O schemes

Scheme	nz	Time (s)		#nit	#lit
lin.TPFA	229 376	205.67	1.0x	653	15 896
nonl.TPFA	367 868	343.75	1.67x	664	23 924
MPFA-O	893 632	833.64	4.05x	660	26 288

3.3.3 Computational Complexity

Different methods of flux approximation result in different computational works. The latter depends on sparsity of the Jacobian matrix, convergence of Newton iterations, and convergence of the linear solver. We compare the linear TPFA, the nonlinear TPFA, and the conventional O-scheme for MPFA.

We consider the simulation of waterflooding process on slightly skewed non-orthogonal $32 \times 32 \times 8$ grid produced from the orthogonal grid by shifting grid nodes by $\Delta z = 0.01\, x + 0.02\, y$. Tensor $\mathbb{K} = \mathbb{R}_z(-\theta)$ diag(1000, ‘10, 10) $\mathbb{R}_z(\theta)$, $\theta = 30°$.

Table 3.3 presents the number of non-zero elements in the Jacobian matrix, CPU time and the total number of nonlinear and linear iterations for 200 days of simulation using linear TPFA, nonlinear TPFA, and MPFA-O methods. Boundary face unknowns are taken into account in the system in the simplest TPFA framework for all three discretization methods. The linear solver is ILU(1)-preconditioned BiCGStab iterations with termination threshold 10^{-12}, and the Newton iterations terminate if the residual is less than 10^{-5}.

The nonlinear TPFA simulation is more expensive than the linear TPFA simulation due to less sparse Jacobian and greater number of accumulated BiCGStab iterations. On the other hand, FV method with the nonlinear TPFA is two fourfold faster than that with the linear MPFA-O. We should note that the linear TPFA provides wrong solution due to the lack of approximation on skewed meshes.

3.3.4 Discrete Maximum Principle

We consider two pseudo-2D numerical experiments, each using three different flux discretization schemes: nonlinear multi-point, nonlinear two-point, and linear two-point. Both test cases use the uniform $32 \times 32 \times 1$ grid for the computational domain $[-50; 50] \times [-50; 50] \times [4000; 4010]$. The first experiment is for one injecting and one producing well, and the second one is for three injecting and one producing wells.

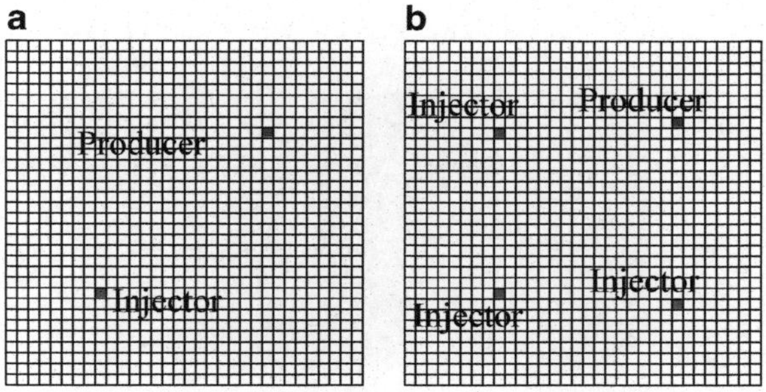

Fig. 3.10 Schemes of the numerical experiments

Well locations are presented in Fig. 3.10. Injecting well pressure is equal to 4100, and producing well pressure is 3900.

Both experiments are conducted for incompressible phases with constant viscosities $\mu_w = 1$, $\mu_o = 50$, constant porosity $\varphi = 0.2$, zero capillary pressure, and zero gravity terms. Relative permeabilities are similar to the first test and the absolute permeability tensor is

$$\mathbb{K} = \mathbb{R}_z(-\theta_z)\text{diag}(k_1, k_2, k_3)\mathbb{R}_z(\theta_z),$$

where $k_1 = k_3 = 100$, $k_2 = 0.1$, $\theta_z = 112.5°$.

Pressure for the two-well problem (Fig. 3.10, left) is presented in Fig. 3.11. The nonlinear multi-point scheme and the linear two-point scheme satisfy the discrete maximum principle for the pressure as expected, while the nonlinear two-point flux discretization violates the discrete maximum principle. It was demonstrated in [51] that the mesh refinement reduces the magnitude of the overshoots and undershoots for the nonlinear two-point scheme, yet the DMP is still violated. In contrast to two nonlinear schemes, the linear two-point scheme does not provide approximation for the fluxes which results in non-physical pressure field: the solution if not aligned with the main anisotropy direction. Therefore, only the nonlinear multi-point scheme is consistent and satisfies the discrete maximum principle.

The four-well test case (Fig. 3.10, right) demonstrates the similar pressure solution violating the discrete maximum principle and the consequences of pressure DMP violation for the saturation solution. Figure 3.12 shows the pressure solution for the time $t = 100$. Again, the discrete maximum principle is violated by the two-point nonlinear scheme in the vast region. The top left injecting well cell has the overshoot, and the pressure is higher than the well bottom hole pressure 4100. As a result, the top left injecting well starts to pump out water instead of injecting. The non-physical flow can be magnified even more if we use the downstream discretization of well mobility $\lambda_{inj} = (\lambda_w + \lambda_o)_{cell} > 0$, and water saturation of the well cell decreases even after it becomes zero. The water saturations obtained by the nonlinear MPFA, the nonlinear TPFA, and the linear TPFA are compared in Fig. 3.13. Due to the lack of the flux approximation of the linear TPFA, the water flow is not alighted with the main anisotropy direction.

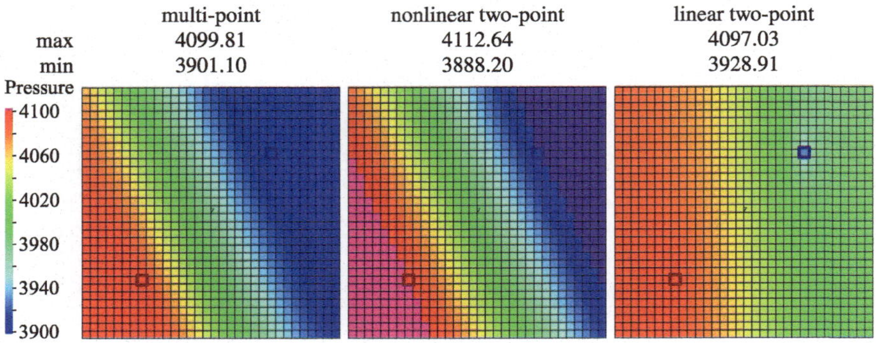

Fig. 3.11 Pressure for $t = 2000$ for different flux discretization schemes (experiment with two wells). Overshoots and undershoots are given by pink and purple colors, respectively

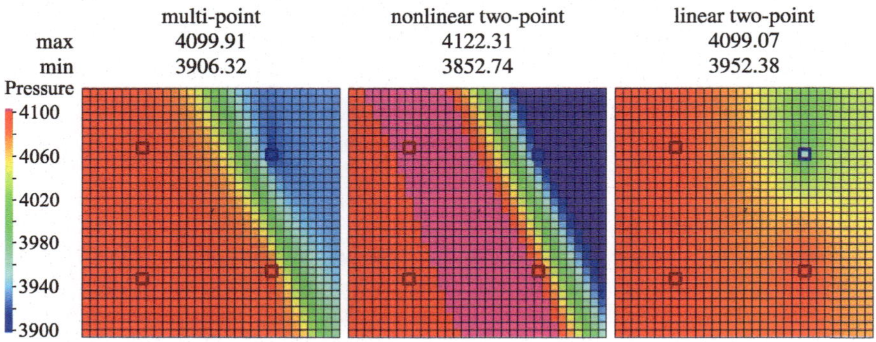

Fig. 3.12 Pressure for $t = 100$ for different flux discretization schemes (experiment with four wells)

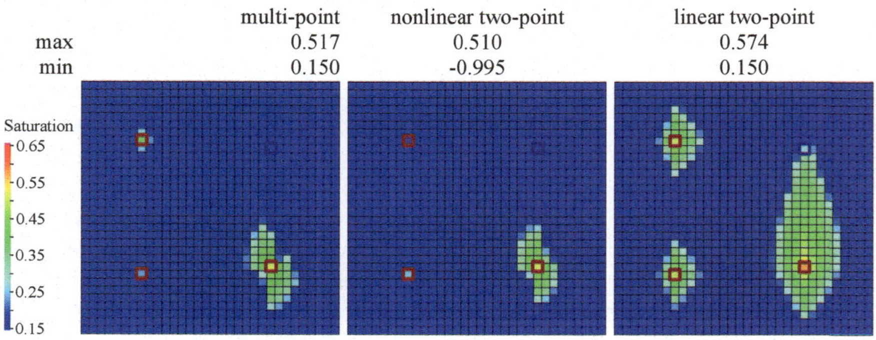

Fig. 3.13 Water saturation for $t = 100$ for different flux discretization schemes (experiment with four wells). Initial saturation is $s(0) = 0.15$

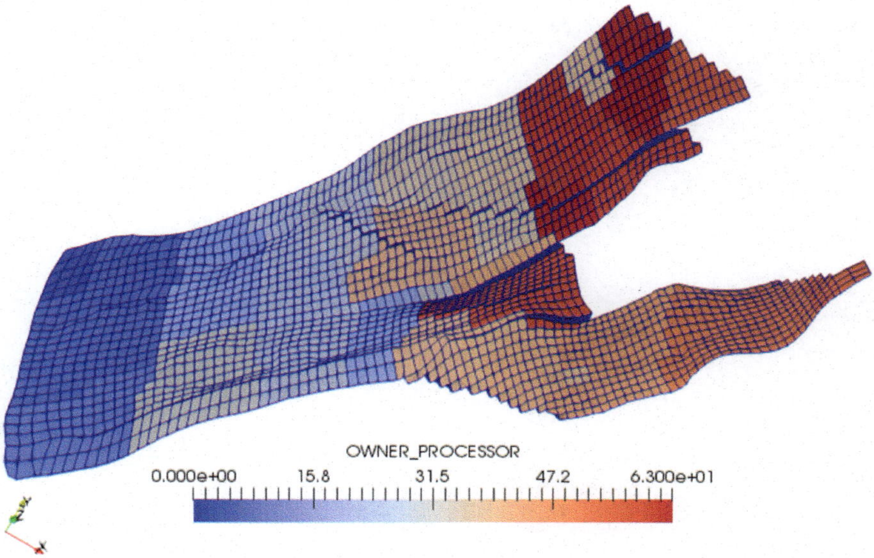

Fig. 3.14 Partitioning of the computational mesh among processors, the black-oil model on the Norne field

3.3.5 Parallel Simulation on the Norne Field

For the black-oil model, we use the secondary recovery problem on the Norne field [8] with one injection well and two production wells. The problem runs for 100 modeling days with at maximum 1-day step. The partition of the mesh is demonstrated in Fig. 3.14. The parameters are taken from SPE9 test.

The results are demonstrated in Table 3.4. The black-oil problem contains a strong elliptic component, and the performance of linear solver deteriorates faster. For good performance, this problem requires a specific solver [99] that can extract elliptic part of the problem and apply a multi-grid solver for it. In this problem, the assembly of the matrix does not ideally scale, since we have to perform certain operations on overlap which may become significant for large number of processors. Still, the simulator demonstrates acceptable parallel speed-up.

3.4 Flow in Fractured Media

The section is devoted to application of the nonlinear FV schemes for flows in fractured porous media [115]. We use the FV method in combination with the embedded discrete fracture method (EDFM), which allows us to recover the flow without mesh adaptation toward fractures. The resulting monotone embedded discrete fracture method (mEDFM) combines effectiveness and simplicity of the standard EDFM

Table 3.4 Performance of the models on different numbers of processors

Two-phase model							
Processors	Assembly		Solution		Total		Equations
1	1028	–	1544	–	2581	–	89830
8	172	6x	383	4x	558	5x	11229
16	89	12x	255	6x	345	8x	5615
32	53	20x	135	12x	189	14x	2808
64	30	35x	71	22x	101	26x	1404
128	19	55x	65	24x	84	31x	702
Three-phase model							
Processors	Assembly		Solution		Total		Equations
1	2783	–	5368	–	8171	–	134745
8	449	6x	3295	2x	3749	2x	16844
16	252	11x	1656	3x	1911	4x	8422
32	152	19x	472	11x	626	13x	4211
64	88	32x	325	17x	415	20x	2106
128	59	47x	154	35x	213	38x	1053

approach with accuracy and physical relevance of the nonlinear FV schemes for non-orthogonal grids and anisotropic media.

The equations for the diffusion problem in fractured porous media with unknown pressure p can be written as

$$\begin{aligned} \big[-\operatorname{div}\mathbb{K}\nabla p\big]^M &= g^M + q^{MF} \quad \text{in } \Omega^M \subset \mathbb{R}^3, \\ \big[p\big]^M &= g_D \qquad\quad \text{on } \Gamma_D^M, \\ \big[\mathbb{K}\nabla p\big]^M \cdot \mathbf{n} &= g_N \qquad\quad \text{on } \Gamma_N^M, \end{aligned} \tag{3.35}$$

for the matrix (i.e., porous media separated from large fractures, denoted by $[\,\cdot\,]^M$) and

$$\begin{aligned} \big[-\operatorname{div}\mathbb{K}\nabla p\big]^F &= g^F + q^{FM} \quad \text{in } \Omega^F \subset \mathbb{R}^2 \\ \big[p\big]^F &= g_D \qquad\quad \text{on } \Gamma_D^F, \\ \big[\mathbb{K}\nabla p\big]^F \cdot \mathbf{n} &= g_N \qquad\quad \text{on } \Gamma_N^F, \end{aligned} \tag{3.36}$$

for the fracture (denoted by $[\,\cdot\,]^F$) spatial domains. Here, Ω^M is a 3D domain with boundary $\Gamma^M = \Gamma_D^M \cup \Gamma_N^M$, $\Gamma_D^M = \overline{\Gamma}_D^M$, $\Gamma_D^M \neq \emptyset$, Ω^F is a 2D domain with boundary $\Gamma^F = \Gamma_D^F \cup \Gamma_N^F$, where $\Gamma_D^F \subset \Gamma_D^M$ and $\Gamma_N^F \subset \Gamma_N^M$ are the corresponding parts of the boundary for the matrix. Besides, $\mathbb{K}$ is a symmetric positive-definite heterogeneous (possibly anisotropic) diffusion tensor, g^M and g^F are source terms in matrix and fractures, respectively, and g_D and g_N are Dirichlet and Neumann boundary conditions, respectively. Finally, $q^{MF} = -q^{FM}$ are the transfer terms between the fracture and the matrix.

3.4.1 Embedded Discrete Fracture Method

The main advantage of EDFM is its ability to account accurately fractures using additional degrees of freedom on meshes not fitted to fractures. The method has obtained close attention in the last few years [77, 103, 112, 141]).

Fractures are assumed to be planar 2D regions with fixed geometry and thickness. Fractures are represented by consistent triangulations. In order to compute parameters for the fracture–cell intersection, each cell is divided into tetrahedra which are intersected by the fracture triangles separately and are used to compute geometric characteristics of the fracture within the cell. Since the fracture geometry is fixed, it is sufficient to compute this data and transmissibility indices once at the model setup stage.

We consider long fractures with lengths $L_F \gg h$ (h is the mesh size in the porous media) since fractures of smaller length can be accounted by modifying local permeability of mesh cells. The fracture width w_F is assumed to be less than h, so one cannot resolve fractures on the reservoir mesh.

The EDFM for the diffusion model (linearized equation (3.1)) is formulated as follows. We introduce two separate domains for the fracture and the medium with corresponding unknown pressures p^F and p^M. We assume that $\mathbb{K}^F = k^F \mathbb{I}$, where $\mathbb{I}$ is the identity tensor. Application of the finite volume method for the coupled fracture–matrix system (3.35)–(3.36) requires three types of fluxes (see Fig. 3.15): cell-to-cell, cell-to-fracture, and intra-fracture exchanges. For the case of several intersecting fractures, we also add fracture-to-fracture fluxes.

Consider a polyhedral mesh Ω_h and a cell $\omega \in \Omega_h$. Let n_ω be the number of fractures $F_i = \Omega_i^F$, $i = 1, .., n_\omega$ crossing ω. Then, the EDFM for the Darcy fluxes in the cell ω can be formulated as follows:

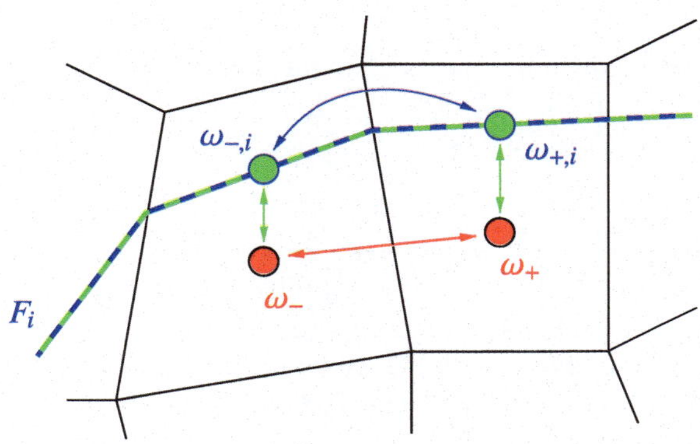

Fig. 3.15 Darcy fluxes for a fracture in porous media: cell-to-cell (red), cell-to-fracture (green), and intra-fracture (blue) exchanges

$$\sum_{f\in\mathcal{F}(\omega)} \mathbf{q}_f\cdot\mathbf{n}_f - \sum_{i=1,\ldots,n_\omega} q_{F_i,\omega} = \int_\omega g^M \mathrm{d}\omega, \tag{3.37}$$

$$\sum_{e_j\in\mathcal{E}(\omega_\rangle)} q_{F_i,e_j} + \sum_{j=1,\ldots,n_\omega} q_{F_{i,j},\omega} + q_{F_i,\omega} = \int_{\omega_i} g^F \mathrm{d}\omega, \qquad i=1,\ldots,n_\omega, \tag{3.38}$$

where $\mathbf{q}_f\cdot\mathbf{n}_f$ is the cell-to-cell diffusive flux between the cell ω and its neighbor through face f and $q_{F_i,\omega}$ is the cell-to-fracture flux from cell ω to fracture F_i. Equation (3.38) is written for each virtual fracture cell $\omega_i = F_i \cap \omega$ with intra-fracture fluxes q_{F_i,e_j} through virtual edges e_j of ω_i. Also Eq. (3.38) accounts for the fracture-to-fracture flux $q_{F_{i,j},\omega}$ between the fracture cell ω_i and other fracture cells $\omega_j = F_j \cap \omega$. It should be noted that $q_{F_{i,j},\omega} = -q_{F_{j,i},\omega}$.

In the monotone EDFM, we use the nonlinear FV schemes for the cell-to-cell fluxes in (3.37): nonlinear TPFA [51] or nonlinear MPFA satisfying the discrete maximum principle (DMP) [48, 143], which were presented in Chap. 2. For the sake of comparison, we will also use two conventional linear schemes: linear TPFA and linear MPFA-O schemes [17].

The general form of the flux discretization will be

$$\mathbf{q}_f\cdot\mathbf{n}_f = T_+\,p_+ - T_-\,p_- \qquad \text{for NTPFA,} \tag{3.39}$$

$$\begin{aligned}\mathbf{q}_f\cdot\mathbf{n}_f &= T_{-,1}\,(p_- - p_{-,1}) + T_{-,2}\,(p_- - p_{-,2}) + T_{-,3}\,(p_- - p_{-,3}) = \\ &= T_{+,1}\,(p_+ - p_{+,1}) + T_{+,2}\,(p_+ - p_{+,2} + T_{+,3}\,(p_+ - p_{+,3}) \qquad \text{for NMPFA,}\end{aligned} \tag{3.40}$$

where coefficients $T_* = T_*(p^M)$ may depend on the unknown in neighboring cells.

For each fracture F_i crossing cells ω_+ and ω_- with common face f, we use a 2D FV scheme for the fracture surface mesh with virtual cells $\omega_{+,i}$ and $\omega_{-,i}$ belonging to ω_+ and ω_-. The FV scheme uses collocation of degrees of freedom at the centers of the virtual cells.

The Darcy flux between fracture cells $\omega_{+,i}$ and $\omega_{-,i}$ on their shared edge e is

$$q_{F_i,e} = \lambda_{F_i,e}(p^F_{-,i} - p^F_{+,i}), \tag{3.41}$$

where the transmissibility in the isotropic media of the fracture is

$$\lambda_{F_i,e} = \frac{k_i^F\, s\, w_i}{a_+ - a_-}, \tag{3.42}$$

Here w_i is the fracture width, s is the length of e (the fracture intersection with the face f), and a_+, a_- are the distances from virtual fracture cell centroids to the edge between $\omega_{+,i}$ and $\omega_{-,i}$ (see Fig. 3.16).

Note that though the tensor in fractures is isotropic, the virtual fracture cells in general case may be non-orthogonal, which will result in 2D linear TPFA scheme not providing approximation. In this case, the nonlinear schemes can be applied for the

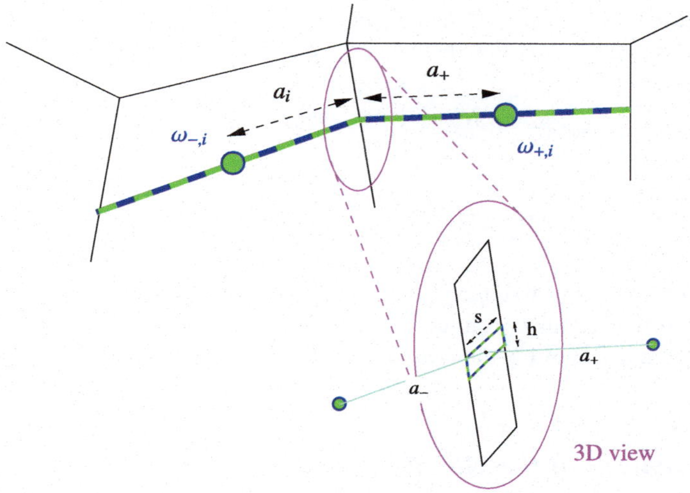

Fig. 3.16 Flux approximation in fracture

intra-fracture fluxes. For the sake of simplicity, we will use the conventional linear TPFA here.

For each cell ω and fracture F_i, the diffusive flux between the fracture and the matrix is

$$q_{F_i,\omega} = \lambda_{F_i,\omega}(p_\omega - p^F_{\omega,i}). \tag{3.43}$$

In order to compute the transmissibility $\lambda_{F_i,\omega}$, we use the transport index approach suggested in [103]. The transmissibility index depends on grid geometry and media physical properties which are known during the simulation. One possible approach for its calculation is to use the harmonic average of the fracture λ_{F_i} and medium λ_ω transmissibilities:

$$\lambda_{F_i,\omega} = \frac{2\,\lambda_{F_i}\lambda_\omega}{\lambda_{F_i} + \lambda_\omega},$$

where

$$\lambda_\omega = \frac{A}{< d >_{\omega,F_i}} \left(\mathbf{n} \cdot \mathbb{K}^M_\omega \mathbf{n}\right), \quad \lambda_{F_i} = \frac{A}{w_i} k^F_i .$$

Here k^F_i is the isotropic permeability within fracture F_i, A is the fracture surface area inside cell ω, $\mathbf{n}$ is the normal vector to fracture inside ω, and $< d >_{\omega,F_i}$ is the averaged normal distance from the fracture to the cell [103]:

$$< d >_{\omega,F_i} = \frac{\int_\omega d_{F_i}(\mathbf{x}) d\mathbf{x}'}{|\omega|},$$

where d_{F_i} is the distance from $\mathbf{x}$ to the fracture and $|\omega|$ is the cell volume.

Intersection of fractures F_i and F_j is a segment which can cross several mesh cells. Within each cell ω, flow between the intersecting fractures can be computed as follows [112]:

$$q_{F_{i,j},\omega} = \lambda_{F_{i,j},\omega}(p_i^F - p_j^F) \tag{3.44}$$

with

$$\lambda_{F_{i,j},\omega} = \frac{\lambda_{F_i}\lambda_{F_j}}{\lambda_{F_i} + \lambda_{F_j}}, \qquad \lambda_{F_i} = \frac{k_i^F w_i s}{a}.$$

Here w_i is the width of fracture F_i, s is the segment intersection length, and a is the averaged normal distance from the center of the fracture subsegments inside cell ω (located at each side of the intersection line) to the intersection line.

3.4.2 Analysis of the Monotone EDFM

In order to solve nonlinear systems produced by the nonlinear discretization schemes and EDFM (3.37)–(3.38), we use either Picard or Newton method. The following statements may be proved for the diffusion equation (the single-phase pressure equation):

Theorem 3.1 *Let the monotone (positivity-preserving) nonlinear TPFA scheme be used for the inter-cell Darcy flux discretization. Then the monotone EDFM with Darcy fluxes* (3.37)–(3.38) *is also positivity-preserving.*

Theorem 3.2 *Let the extremum-preserving nonlinear MPFA scheme be used for the inter-cell Darcy flux discretization. Then the monotone EDFM with Darcy fluxes* (3.37)–(3.38) *is also extremum-preserving.*

The proofs [115] are based on the properties of a linear system which appears for each Picard iteration. As soon as the Picard method converges, the resulting solution will have desired positivity-preserving or extremum-preserving properties. In practice, one can apply the Newton method which converges faster but requires construction of the Jacobian matrix.

3.4.3 Numerical Experiments

In this section, we consider two numerical experiments for flows in fractured media:

- The first experiment is designed to study the ability of the mEDFM to satisfy the discrete maximum principle (DMP) for different discretization schemes for (3.35). We consider the diffusion problem from Chap. 2 where several highly permeable fractures are added.

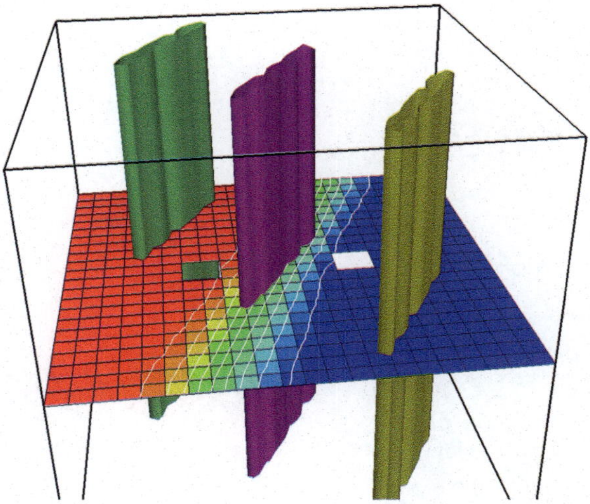

Fig. 3.17 DMP test: fractures' location

- The second experiment is an unsteady two-phase flow waterflood test with several wells and a fracture. The mEDFM is compared with the discrete fracture method (DFM) which uses a modified grid resolving the fracture.

The first test studies the DMP features of the discretization schemes. The domain is the unit cube without two boxes of imitating wells: $\Omega = [0, 1]^3 \setminus (\Omega_1 \cup \Omega_2)$, $\Omega_1 = [3/11, 4/11] \times [5/11, 6/11] \times (0, 1)$, $\Omega_2 = [7/11, 8/11] \times [5/11, 6/11] \times (0, 1)$.

The test setting repeats the second numerical test from Chap. 2 (Fig. 2.3), where we add three vertical rectangular fractures with the following corner points (Fig. 3.17):

$A_1 = (0.16, 0.3268, 0)$, $B_1 = (0.36, 0.6732, 1)$,
$A_2 = (0.41, 0.3268, 0)$, $B_2 = (0.61, 0.6732, 1)$,
$A_3 = (0.66, 0.3268, 0)$, $B_3 = (0.86, 0.6732, 1)$.

Fractures' locations are chosen to test possible DMP violations. Fractures' width is $w_{f,i} = 0.01$, and permeabilities in fractures are isotropic: $k_i^F = k^F = 1000$, $i = 1, 2, 3$.

Table 3.5 shows minima and maxima of the FV solution by TPFA, MPFA-O, nonlinear TPFA (positive), and nonlinear MPFA (satisfying DMP) schemes, and the mEDFM is applied to account for the fractures. The mEDFM with linear TPFA, as expected, provides no approximation but preserves maximum and minimum of the discrete solution. Both MPFA-O and nonlinear TPFA discretizations violate the DMP, while MPFA-O discretization also violates solution positivity. The schemes violating the DMP generate undershoots and overshoots even in fractures, despite the fact that the discretization scheme for (3.36) is the linear TPFA satisfying the DMP. Only the nonlinear MPFA preserves both maximum and minimum while showing reasonable solution. Figure 3.18 presents the FV solutions of the diffusion problem on a mesh cross section.

Table 3.5 Minima and maxima of the FV solution in matrix and fractures for the DMP test. Overshoots and undershoots are marked by *

	$min(p)$	$max(p)$	$min(p_f)$	$max(p_f)$
mEDFM (TPFA)	0.0245	0.9755	0.1376	0.8534
mEDFM (NTPFA)	0.0063	1.7395*	0.1131	1.3636*
mEDFM (NMPFA)	0.0074	0.9925	0.0244	0.9750
mEDFM (MPFA-O)	–0.0459*	1.0442*	–0.0015*	0.9995

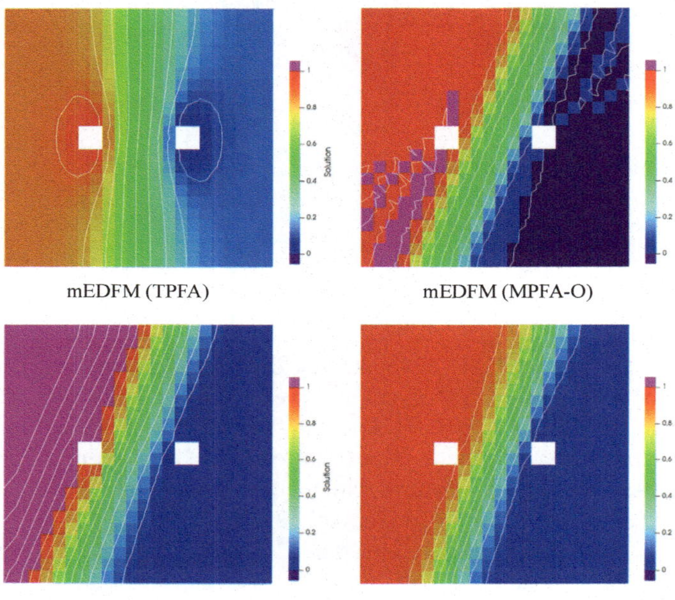

Fig. 3.18 DMP test: FV solutions for the domain with three fractures (overshoots and undershoots are colored in pink and dark blue, respectively)

The second test addresses the two-phase waterflood for a standard five-spot problem with two wells in the opposite corners of a rectangular domain, where we add two fractures as shown in Fig. 3.19.

The permeability tensor for the porous media is full and anisotropic:

$$\mathbb{K}^M = \mathbb{R}_z(-\alpha)\begin{pmatrix} k_1 & 0 & 0 \\ 0 & k_2 & 0 \\ 0 & 0 & k_3 \end{pmatrix}\mathbb{R}_z(\alpha), \qquad \mathbb{R}_z(\alpha) = \begin{pmatrix} \cos\alpha & \sin\alpha & 0 \\ -\sin\alpha & \cos\alpha & 0 \\ 0 & 0 & 1 \end{pmatrix},$$

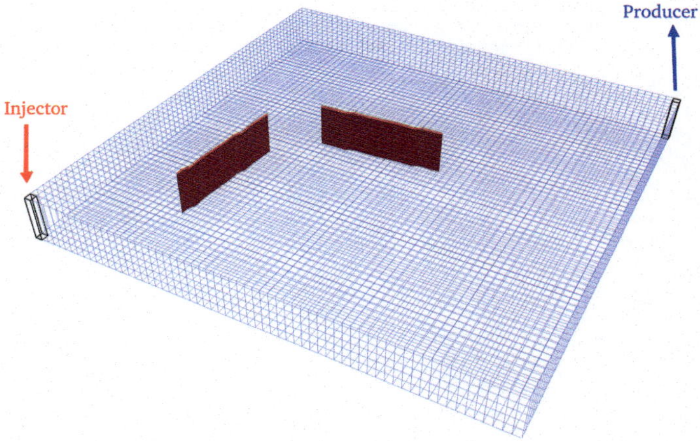

Fig. 3.19 Setup for the five-spot problem with two fractures

where $k_1 = 10^3$ [md], $k_2 = k_3 = 10^2$ [md], $\alpha = \frac{\pi}{4}$, and the porosity is $\phi^M = 0.15$.

The permeability tensor for the fractures is scalar $\mathbb{K}^F = k^F \mathbb{I}$, $k^F = 10^6$ [md], $w_F = 0.13$ [ft], and the porosity is $\phi^F = 0.15$.

Domain dimensions are $[0, 100] \times [0, 100] \times [0, 10]$ ft. Tables for capillary pressure and relative permeabilities are similar to the two-phase flow experiments from Sect. 3.3. For the injector and producer wells, we set the bottom hole pressures $p_{inj} = 4100$ [psi] and $p_{prod} = 3900$ [psi]. The initial pressure is $p_0 = 4000$ [psi], and the initial saturation is $S_0 = 0.15$.

We simulate the water flood for 90 days with time step $\Delta t = 1$ day and compare three solutions: (1) the EDFM solution with the linear TPFA discretization for all flux types, (2) the mEDFM solution with the nonlinear TPFA discretization between cells, and (3) the discrete fracture method (DFM-FV) solution with the nonlinear TPFA scheme, which applies directly to the original FV discretization for the mesh with cut cells and a thin layer of 3D cells representing the fracture.

The water and oil rates for the producer well are shown in Fig. 3.20. The mEDFM and the DFM-FV schemes produce very close results with similar rates and breakthrough times since the NTPFA scheme provides the approximation for non-$\mathbb{K}$-orthogonal grids. On the contrast, the original EDFM provides a different solution, with 40% larger breakthrough time.

Figure 3.21 shows the oil pressure and the water saturation fields at the time $T = 45$ days. One can see that the mEDFM (NTPFA) and the DFM-FV methods produce almost identical results, whereas the EDFM solution is noticeably different from them. The DFM-FV requires grid modification to take fractures into account explicitly, which may complicate the reservoir simulation. The mEDFM provides a viable alternative.

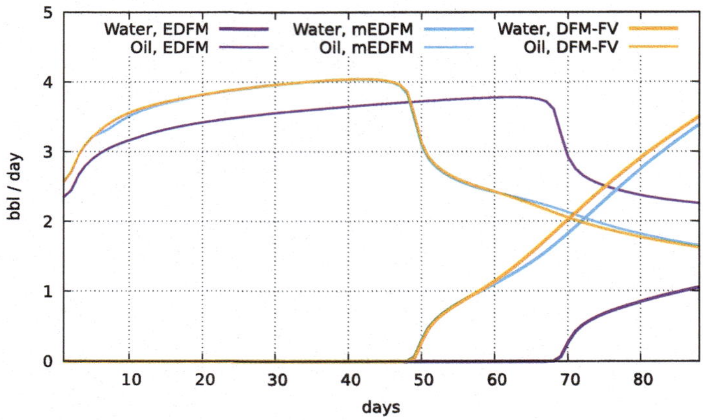

Fig. 3.20 Oil and water rates for EDFM, mEDFM (NTPFA), and DFM-FV (NTPFA) solutions

3.5 Near-Well Correction Method

In this section, we present a possible alternative to the Peaceman well model, which was discussed in Sect. 3.1.4. The near-well correction (NWC) method takes into account the logarithmic singularity of the pressure in a near-well region [56, 57] and introduces a correction to improve accuracy of the pressure and the flux computation.

Consider a near-well region which spans well singularity (see Fig. 3.22). We modify the nonlinear monotone FV scheme and take into account the solution singularity near an isolated well. The method is designed for anisotropic media, arbitrary polyhedral cells, and arbitrary well location.

In the nonlinear FV method presented in Sect. 2.4, the discrete fluxes are calculated on the basis of the piecewise linear reconstruction of the unknown field. The NWC method takes into account the nonlinearity of the solution near the well.

We consider the pressure field to be the sum of linear and nonlinear functions for each cell in a near-well region:

$$p_\omega = \underbrace{a\,x + b\,y + c\,z + d}_{p_{lin}} + \underbrace{e\,\mathcal{S}(x, y, z)}_{p_{\mathcal{S}}}, \tag{3.45}$$

where $\mathcal{S}(x, y, z)$ is a function representing the singularity.

The finite volume discretization requires the normal component of the flux $\mathbf{q} = -\mathbb{K}\nabla p$ to be integrated on each face f of ω:

$$\int_f \mathbf{q}\cdot\mathbf{n}_f \mathrm{d}s = -\int_f (\mathbb{K}\nabla p_\omega)\cdot\mathbf{n}_f \mathrm{d}s = -\int_f (\mathbb{K}\nabla p_{lin})\cdot\mathbf{n}_f \mathrm{d}s - \int_f (\mathbb{K}\nabla p_{\mathcal{S}})\cdot\mathbf{n}_f \mathrm{d}s. \tag{3.46}$$

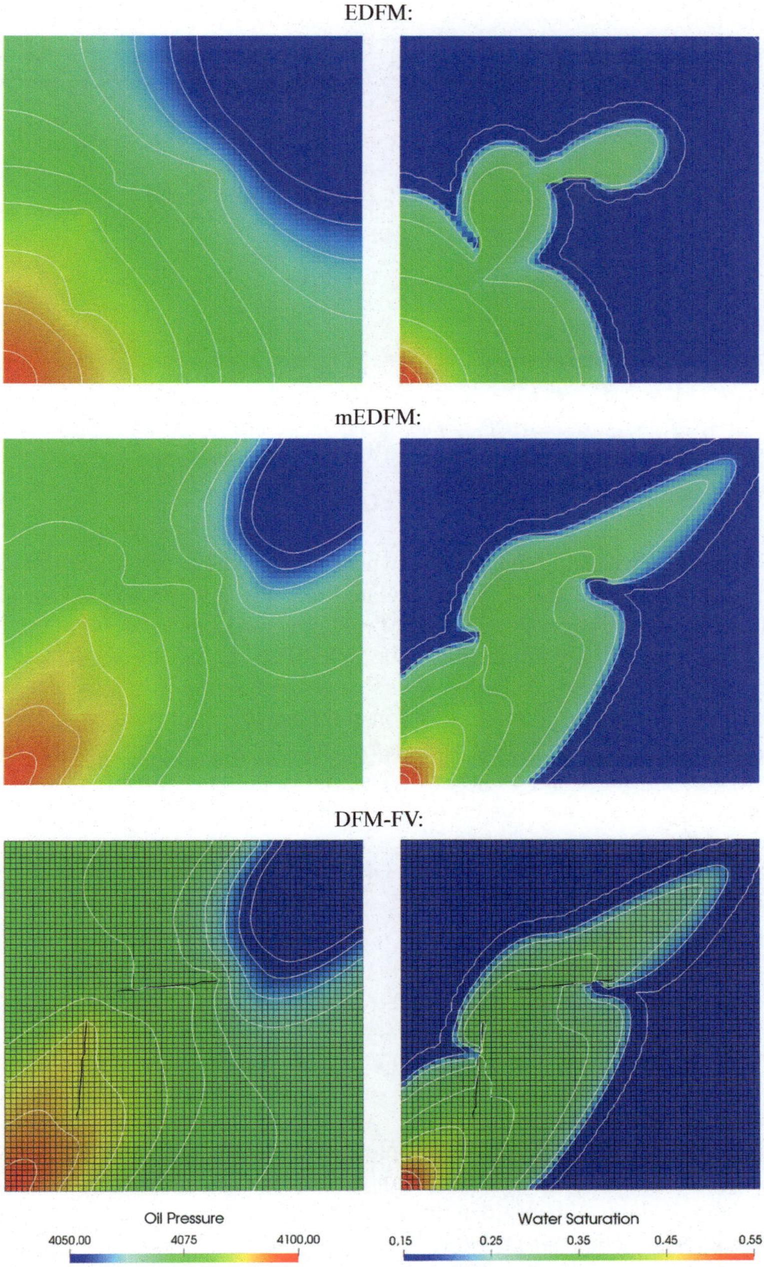

Fig. 3.21 Oil pressure (left) and water saturation (right) fields for the two-phase flow, T = 45 days. Top: EDFM solution; middle: mEDFM (NTPFA) solution; bottom: DFM-FV (NTPFA) solution

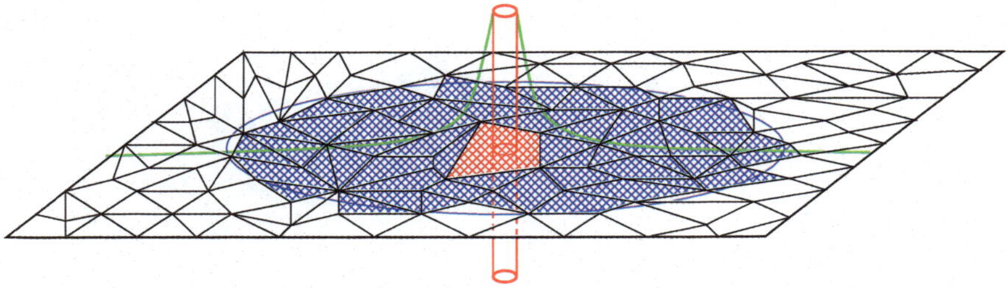

Fig. 3.22 Logarithmic singularity in the near-well region

Combining (3.45) and (3.46) yields the mean normal flux

$$q_f = \int_f \mathbf{q} \cdot \mathbf{n}_f \mathrm{d}s = -\int_f \mathbb{K} \begin{pmatrix} a \\ b \\ c \end{pmatrix} \cdot \mathbf{n}_f \mathrm{d}s - e \int_f (\mathbb{K} \nabla \mathcal{S}(x, y, z)) \cdot \mathbf{n}_f \mathrm{d}s$$
$$= a\boldsymbol{\ell}_1 + b\boldsymbol{\ell}_2 + c\boldsymbol{\ell}_3 + e\boldsymbol{\ell}_4. \quad (3.47)$$

In the following, we shall omit index f whenever it does not result in confusion.

Integrals for $\boldsymbol{\ell}_1$, $\boldsymbol{\ell}_2$, and $\boldsymbol{\ell}_3$ are calculated exactly. Integral for $\boldsymbol{\ell}_4$ can also be calculated exactly for some simple cases of well, grid, and tensor, but for more general cases the numerical integration should be used. The coefficients $\boldsymbol{\ell}_i$ depend solely on the mesh and problem data, and are calculated explicitly, while the coefficients (a, b, c, e) are recovered from the solution at the neighboring cells.

Let ω_+ and ω_- be neighboring cells sharing a face f, and $\mathbf{x}_+, \mathbf{x}_-$ denote centroids of these cells. We take four points $\mathbf{x}_i$ ($\mathbf{x}_i \neq \mathbf{x}_+$) and call four vectors $\mathbf{t}_i = \mathbf{x}_i - \mathbf{x}_+$ the *quadruplet*. Points $\mathbf{x}_i$ denote centers of the neighboring cells or faces of ω, $p_i = p(\mathbf{x}_i)$ and $p_+ = p(\mathbf{x}_+)$.

We assume the same representation (3.45) for vectors of each quadruplet, which gives us

$$\begin{pmatrix} p_1 - p_+ \\ p_2 - p_+ \\ p_3 - p_+ \\ p_4 - p_+ \end{pmatrix} = \begin{bmatrix} x_1 - x_+ & y_1 - y_+ & z_1 - z_+ & \mathcal{S}_1 - \mathcal{S}_+ \\ x_2 - x_+ & y_2 - y_+ & z_2 - z_+ & \mathcal{S}_2 - \mathcal{S}_+ \\ x_3 - x_+ & y_3 - y_+ & z_3 - z_+ & \mathcal{S}_3 - \mathcal{S}_+ \\ x_4 - x_+ & y_4 - y_+ & z_4 - z_+ & \mathcal{S}_4 - \mathcal{S}_+ \end{bmatrix} \begin{pmatrix} a \\ b \\ c \\ e \end{pmatrix}, \quad (3.48)$$

where $\mathcal{S}_* = \mathcal{S}(x_*, y_*, z_*)$.

The collocation points of the quadruplet should be chosen carefully in order to avoid degenerated matrix in (3.48). Among all admissible quadruplets, we choose the one with the largest matrix determinant.

Solving the system (3.48) provides us the coefficients a_+, b_+, c_+, e_+ for the cell ω_+:

$$a_+ = \sum_j (p_j - p_+)\, m_{1,j}, \qquad b_+ = \sum_j (p_j - p_+)\, m_{2,j},$$
$$c_+ = \sum_j (p_j - p_+)\, m_{3,j}, \qquad e_+ = \sum_j (p_j - p_+)\, m_{4,j}, \tag{3.49}$$

where $m_{i,j}$ are the elements of the inverse matrix from (3.48). Taking ω_- instead of ω_+ and considering $-\mathbf{q} \cdot \mathbf{n}_f$ provides us the second flux approximation.

Applying (3.49) to Eq. (3.47) gives us

$$q_\pm = \pm \int_f \mathbf{q} \cdot \mathbf{n}_f ds = \pm\Big[\boldsymbol{\ell}_1 \sum_j (p_j - p_\pm)\, m^\pm_{1,j} + \boldsymbol{\ell}_2 \sum_j (p_j - p_\pm)\, m^\pm_{2,j} \tag{3.50}$$
$$+ \boldsymbol{\ell}_3 \sum_j (p_j - p_\pm)\, m^\pm_{3,j} + \boldsymbol{\ell}_4 \sum_j (p_j - p_\pm)\, m^\pm_{4,j}\Big], \tag{3.51}$$

or

$$q_\pm = \pm\Big[\sum_j p_j \underbrace{\sum_i \boldsymbol{\ell}_i\, m^\pm_{i,j}}_{k^\pm_j} - p_\pm \sum_j \underbrace{\sum_i \boldsymbol{\ell}_i\, m^\pm_{i,j}}_{k^\pm_j}\Big] = \pm\Big(\sum_j k^\pm_j\, (p_j - p_\pm)\Big). \tag{3.52}$$

The resulting flux approximation is obtained as the weighted sum of q_+ and q_- with coefficients $\mu_+ + \mu_- = 1$

$$q_f = \mu_+ \Big(\sum_j k^+_j\, (p_j - p_+)\Big) - \mu_- \Big(\sum_{j'} k^-_{j'} \cdot (p_{j'} - p_-)\Big). \tag{3.53}$$

To construct the *linear* multi-point flux discretization, we fix $\mu_+ = \mu_- = 1/2$.

The particular choice of singularity function $\mathcal{S}$ in (3.45) depends on geometric and physical assumptions. A perfect well in isotropic media allows us to use the Dupuit formula and set $\mathcal{S}(x, y, z) = ln(r)$, where $r(x, y, z)$ is the distance to the well axis. For more complicated cases of anisotropic media and partially perforated wells, we refer to [98].

The near-well correction method generalizes the conventional Peaceman well model [123] and thus is applicable to the cases of arbitrary polyhedral cells, slanted wells, and wells separated from grid cell centers.

The original Peaceman formula was derived for the perfect vertical well where all the neighboring cell pressures are given by the Dupuit formula to catch the logarithmic behavior of the solution in the near-well region. Also, it is assumed that the well flux is balanced by the sum of the *linear* TPFA for the well cell faces. These assumptions require selection of an equivalent radius, which is used for definition of the well cell pressure and ensures flux continuity.

In contrast to the Peaceman approach, the new well model is incorporated in the near-well correction scheme which takes the singularity into account by construction

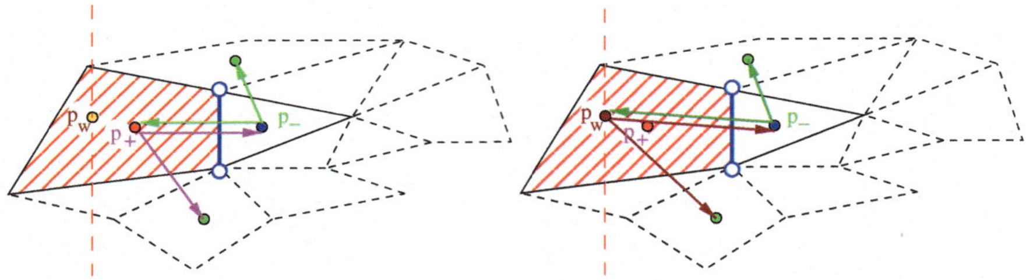

Fig. 3.23 Well cell: stencils for the reservoir pressure (left) and the additional well pressure (right)

and does not impose additional restrictions on the well cell degree of freedom. For each well cell, we introduce an additional point on the well segment associated with the bottom hole pressure. Using this point and considering only outer flux ($\mu_+ = 1$, $\mu_- = 0$) in (3.53), we get an additional relation for the facial fluxes of the well cell. Therefore, for the well cell faces we have two flux approximations with different stencils (see Fig. 3.23). For each quadruplet calculation, the points inside the well are projected to the well surface in order to avoid the singularity.

Summing fluxes for the well cell gives us the well cell equation:

$$\sum_f \left[\frac{1}{2} \Big(\sum_j k_j^+ \left(p_j - p_+\right) \Big) - \frac{1}{2} \Big(\sum_{j'} k_{j'}^- \left(p_{j'} - p_-\right) \Big) \right]$$
$$= \sum_f \left[1 \Big(\sum_l k_l^+ \left(p_l - p_w\right) \Big) - 0 \Big(... \Big) \right]. \tag{3.54}$$

If we use the given well flux q_w condition, the additional equation for unknown p_w will occur:

$$\sum_{\text{well cells}} \sum_f \Big(\sum_l k_l^+ \left(p_l - p_w\right) \Big) = q_w. \tag{3.55}$$

3.5.1 Numerical Experiment

We study the NWC scheme on an analytical solution which is known for two vertical wells in the box domain. Domain dimensions are $[-100; 100] \times [-50; 50] \times [0; d]$. We consider the domain as pseudo-2D and neglect z coordinate in further description. The wells are located at $(-50, 0)$ and $(50, 0)$, and the well rates are $q_1 = 1$ and $q_2 = 4$. The permeability tensor is scalar, $\mathbb{K} = \mathbb{I}$.

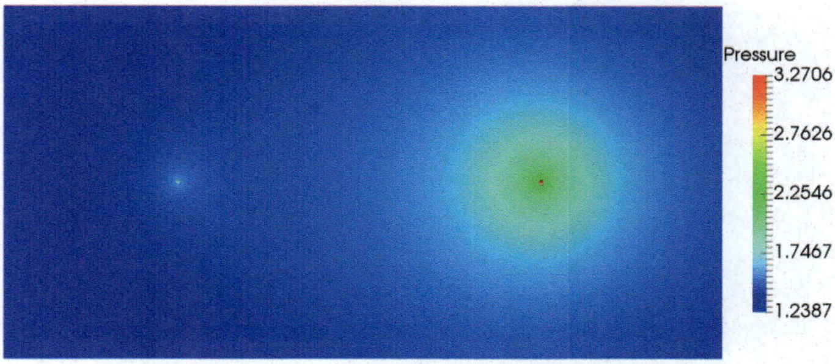

Fig. 3.24 Analytical solution for two-well problem

Table 3.6 Solution relative errors for p and flux errors for q_1 and q_2 for the problem with two wells for cubic grids

$100/h$	$\varepsilon_{p,pcm}^{NFV}$	ε_{p}^{NWC}	$\varepsilon_{q_1}^{NFV}$	$\varepsilon_{q_2}^{NFV}$	$\varepsilon_{q_1}^{NWC}$	$\varepsilon_{q_2}^{NWC}$
33	1.2e-2	2.8e-5	4.6e-3	1.9e-2	2.1e-5	4.1e-5
67	5.2e-3	7.6e-6	4.6e-3	1.9e-2	2.3e-5	5.4e-5
99	3.1e-3	4.1e-6	4.6e-3	1.8e-2	2.0e-5	7.0e-5

In order to fix a unique solution, we set the pressure in the middle point $\mathbf{x}_0 = (0, 0)$, $P_0 = p(\mathbf{x}_0) = 1.5$. The analytical solution [76] is shown in Fig. 3.24:

$$p(\mathbf{x}) = P_0 - \frac{q_1 \ln(r_1/r_{w,p1})}{2\pi k h_w} + \frac{q_2 \ln(r_2/r_{w,p2})}{2\pi k h_w},$$

where h_w is the well height ($h_w = d = 3$ in our case), r_1, r_2 are the distances from point $\mathbf{x}$ to the wells 1 and 2, respectively, and $r_{w,p1}$, $r_{w,p2}$ are the distances from the middle point $\mathbf{x}_0$ to the wells. Pressures on the wells are obtained from this formula.

The test uses the simplest cubic grids which are the best for the Peaceman method. Grid dimensions are $66 \times 33 \times 1$, $134 \times 67 \times 1$ and $198 \times 99 \times 1$. The radii of the near-well regions used in the logarithmic correction for both wells are $R_1 = R_2 = 30$.

The Peaceman formula is used with the nonlinear monotone scheme (2.14) [113], while the new well cell model is used in combination with the NWC scheme. We compute both relative L^2 error norms for the pressure ($\varepsilon_{p,pcm}^{NFV}$ and ε_{p}^{NWC}) and the well rate errors (ε_{q}^{NFV} and ε_{q}^{NWC}).

Table 3.6 shows the relative errors for the NFV and the NWC methods for the analytical well rates, and the relative errors for the pressure and the well rates (the first and the second well) for the numerical well models: NFV + Peaceman and the NWC method.

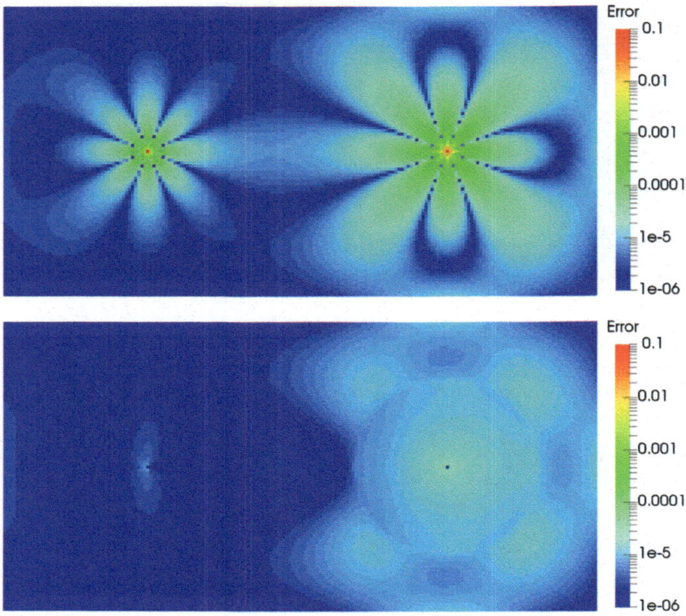

Fig. 3.25 Relative errors for the NFV scheme with Peaceman well model (top) and the NWC (bottom) methods in the log-scale. Cubic grid $134 \times 67 \times 1$

Figure 3.25 presents the error fields for the NFV scheme with the Peaceman well model and the NWC method in the log-scale. Note that the NFV scheme reduces to the standard FV scheme with the linear two-point flux approximation on cubic mesh and isotropic media. The largest error of the NFV scheme is concentrated in regions around the wells that are covered by the near-well regions of the NWC method. The NWC method gives considerably smaller errors than the conventional method.

Chapter 4
Application of FVM in Modeling of Subsurface Radionuclide Migration

In this chapter, the hydrogeological multi-physics models and corresponding numerical methods are presented basing on the authors' experience of GeRa hydrogeological code [87, 93] development and its applications. Flow in unsaturated conditions, reactive transport, and density-driven flow models are addressed.

4.1 Domains, Physics, and Mathematical Models for Subsurface Radionuclide Migration

Safe radioactive waste (RW) disposal is one of the key problems to be solved to allow for nuclear energy sustainable development. The contemporary concepts for RW disposal facilities include surface and subsurface repositories for low- and intermediate-level waste (LILW) and ultimately deep geological repositories (DGR) for high-level waste (HLW) and long-living intermediate-level waste. In both cases, prevention of groundwater pollution is an important task. LILW disposal facilities usually rely mainly on the safety functions of the engineered barrier system (EBS). On the contrary, the safety of HLW deep geological repositories is provided mainly by the geological media because of very long time periods of waste remaining hazardous. Three major types of geological media are considered for DGR displacement: crystalline rock (Finland, Sweden, Russia, Japan), clay (France, Belgium, Canada et al.), and salt (Germany, USA) formations. Safety assessment of RW demands modeling groundwater flow and radionuclides transport in geological media. In case of DGR located in crystalline rock, flow and transport occur in a system of fractures while the rock matrix is almost impermeable but may capture contaminants by means of diffusion and sorption. Modeling hydrogeological processes in crystalline rock usually involves generation of discrete fracture networks (DFNs) and either solving the flow and transport problems directly on DFN or using the equivalent porous media approach via upscaling techniques [75, 102]. Flow and transport processes in clay and salt rocks are normally considered in the continuum approach.

Y. Vassilevski et al., *Parallel Finite Volume Computation on General Meshes*,
https://doi.org/10.1007/978-3-030-47232-0_4

In this chapter, we will describe models and corresponding numerical methods basing on the authors' experience of GeRa hydrogeological code [90, 96] development and its applications. The numerical part of the chapter will concern mainly the temporal discretization and coupling problems, as far as the space discretization is based on linear (two-point and multi-point flux approximation, namely, O-scheme [16]) and nonlinear monotone FV schemes described in Chap. 2.

Here, we should generally formulate the tasks which are usually solved in radioactive waste repository safety assessment. As far as the main goal is to prevent any harm to people and biosphere now and in future caused by radionuclides, one needs to determine the ways how they can reach the biosphere. Then it is necessary to forecast the exposure doses for the inhabitants. In case of long-lived waste, the time periods for the safety assessment are very long. Namely, in many countries one million years is considered as the reference time period. It may be even larger, for example, the Russian legislation demands proof of repository safety for the whole period of its potential hazardness. Even though these time periods are huge with respect to human life and characteristic times of common processes in the biosphere, they can be small compared to the ages of the corresponding geological formations in which the wastes are going to be disposed, providing an extra argument that the rock can maintain its integrity and serve as safe barrier for this long. When making safety assessments for LILW, the time periods for forecasts are much smaller, usually from several hundreds to thousands of years depending on the waste type. In order to make the safety assessment, we need to understand the possible influence of various features, events, and processes. As the predictions for large time periods suffer greatly from uncertainties, their quantification is a compulsory part of the assessment, and the calculation of the worst repository evolution scenario is one of the major tools to eliminate a huge part of uncertainties. Normally, modeling transport of radionuclides from the repository to biosphere involves at least two models. The near-field one includes the waste immobilization matrix, EBS, and excavation damaged zone. The near-field model shall take into account the evolution of the EBS as well as the transport processes. The far-field model which is the focus of this chapter takes the output of the near-field model as a source and simulates transport in geological media. Sometimes, the need for intermediate models arises, for example, to take into account the repository structure [72, 86].

Since the main process affecting transport of radionuclides from a repository is groundwater flow, the best choice of the modeling domain is a catchment. Its boundaries are usually defined as atmospheric boundary on the top surface including rainfall recharge, evaporation, and transpiration by plant roots. The bottom boundary can be considered as impermeable if there is an aquitard below the domain or it can be a prescribed head boundary in case of an aquifer below. The side boundaries should be defined by watersheds, streamlines, or neighboring reservoirs. Reservoirs may also be present on the top. The main processes to be taken into account are the following:

- Groundwater flow in various regimes: confined, unconfined, unsaturated, and two-phase flow of water and air.
- Basic transport processes: advection, diffusion, and dispersion.

- Chemical interactions in the system of solute and rock.
- Radioactive decay with possible decay chains.
- Density-driven flow.
- Heat transport and thermohaline convection.
- Surface flow.
- Geomechanics.

The basic model for radionuclides transport calculation consists of the flow equation (see, for example, Eq. (5.1.55) in [31])

$$s_0 \frac{\partial \mathrm{h}}{\partial t} + \mathrm{div}\mathbf{q} = Q \tag{4.1}$$

and the transport equation

$$R\varphi \frac{\partial c}{\partial t} - \mathrm{div}\, \mathbb{D}\nabla c + \mathrm{div}\, (\mathbf{q}c) + \lambda\varphi Rc = f. \tag{4.2}$$

Here, s_0 is the specific storativity, h is the piezometric head, $\mathbf{q}$ is the volumetric water flux, and Q is the volumetric source term. The flow equation is written in the assumption of constant water density. In the transport equation, R is the retardation factor, φ is the porosity, $\mathbb{D}$ is the diffusion–dispersion tensor, c is the contaminant concentration, λ is the decay constant, and f is the mass source term of the contaminant. To eliminate variable $\mathbf{q}$, Darcy law is used:

$$\mathbf{q} = -\mathbb{K}\nabla \mathrm{h}, \tag{4.3}$$

in which $\mathbb{K}$ is the hydraulic conductivity, a symmetric positive-definite matrix 3×3.

Depending on the isotherm used, the retardation factor may be a constant or a function of concentration. The most widely used is the linear or Henry isotherm stating that the amount of sorbed contaminant is proportional to its concentration in the solute, in this case

$$R = 1 + \frac{\rho_b K_d}{\varphi}, \tag{4.4}$$

K_d being the distribution coefficient and ρ_b the rock bulk density. Among the nonlinear isotherms, the most widely used are the power law Freundlich isotherm [68] and the Langmuir isotherm [101] allowing to take into account finite cation exchange capacity of the medium.

Diffusion–dispersion tensor $\mathbb{D}$ also called coefficient of hydrodynamic dispersion [31] is a 3×3 symmetric positive-definite matrix. In the assumption of isotropic dispersion, it is defined by formula

$$\mathbb{D} = \mathbb{D}_m + \mathbb{D}_d = d_m\mathbb{I} + \|\mathbf{q}\|(\alpha_l\mathbb{E}(\mathbf{q}) + \alpha_t(\mathbb{I} - \mathbb{E}(\mathbf{q}))) \tag{4.5}$$

with $\mathbb{D}_m$ denoting molecular diffusion (first term in the RHS) and $\mathbb{D}_d$ hydrodynamic dispersion (second term in the RHS) contributions; $\mathbb{I}$ being identity matrix of order 3, d_m being effective molecular diffusion coefficient, α_l and α_t being longitudinal and transverse dispersivities and $\mathbb{E}(\mathbf{q})$ being the tensor with components $E_{ij}(\mathbf{q}) = \frac{q_i q_j}{\|\mathbf{q}\|^2}$, $i, j \in \{1, 2, 3\}$.

The decay constant λ is derived from the radionuclide half-life period $T_{1/2}$ using the relation

$$\lambda = \frac{\ln 2}{T_{1/2}}. \tag{4.6}$$

This decay model is used in case of modeling transport of a single radionuclide producing a stable or not hazardous isotope, for example, a very long-living isotope, which will anyway have very low activity. Also, this may be the case when we are not interested in modeling transport of a short-lived (with respect to the parent one) daughter isotope as its activity is approximately equal to the activity of the parent isotope. The aforementioned model is applicable for Strontium-90, Cesium-137, and Iodine-129. When modeling transport of isotopes featuring long decay chains such as Uranium-235, Uranium-238, Thorium-232, Neptunium-237, and their decay and transmutation products, one has to use a model of multicomponent transport with chain decay and production.

The model (4.1)–(4.2) has to be supplied with boundary and initial conditions. For the flow equation, the boundary conditions are usually either the no-flow or fixed-flux conditions (Neumann-type)

$$\mathbf{q} \cdot \mathbf{n}|_{\Gamma_N} = g_N(\mathbf{x}, t), \tag{4.7}$$

fixed piezometric head (Dirichlet-type)

$$\mathrm{h}|_{\Gamma_D} = g_D(\mathbf{x}, t), \tag{4.8}$$

or Robin-type boundary condition setting the relation between the piezometric head and the flux through the boundary

$$\mathbf{q} \cdot \mathbf{n}|_{\Gamma_R} = \alpha\left(H(\mathbf{x}, t) - \mathrm{h}\right)|_{\Gamma_R}, \tag{4.9}$$

where α is the conduction coefficient and $H(\mathbf{x}, t)$ normally denotes the water level in the neighboring reservoir.

For the transport equation, one can set Dirichlet-type boundary conditions prescribing the concentration

$$c|_{\Gamma_D} = c_D(\mathbf{x}, t), \tag{4.10}$$

Neumann-type, prescribing the diffusion flux through the boundary

$$-\mathbb{D}\nabla c \cdot \mathbf{n}|_{\Gamma_N} = q_N(\mathbf{x}, t), \tag{4.11}$$

or the total flux boundary condition in the form

$$(-\mathbb{D}\nabla c + \mathbf{q}c) \cdot \mathbf{n}|_{\Gamma_R} = g_R(\mathbf{x}, t) . \tag{4.12}$$

Neumann-type conditions are usually imposed with zero flux either on impermeable boundaries or on distant boundaries from the contaminant source where one can assume that the diffusion–dispersion flux is negligible. The total flux condition can be used to prescribe the concentration in the infiltration flux.

Initial conditions for the problem comprise initial piezometric head $\mathrm{h}_0(\mathbf{x})$ and concentration $c_0(\mathbf{x})$.

4.2 Flow in Unconfined and Unsaturated Conditions, Transport in Vadose Zone

4.2.1 Mathematical Models

Strictly speaking, the flow and transport model (4.1)–(4.2) may be applied only for fully saturated media. But most often one has to deal with a basin which has a free water table or contains partially saturated areas. The application of two-phase (water–gas) flow models in groundwater flow simulation is limited to specific cases when the variations in gas pressure in the domain are significant. A simplified model of unsaturated flow presuming constant atmospheric pressure in the gas phase is usually preferred. For problems with a free water table also the models of flow in unconfined regime are used. The latter assume that the flow can occur only in the saturated zone.

Flow in unsaturated conditions is governed by the Richards equation written in terms of moisture content [125] and pressure head

$$\frac{\partial \theta}{\partial t} + Ss_0 \frac{\partial \psi}{\partial t} - \operatorname{div}\left(k_r(\theta)\mathbb{K}\nabla(\psi + z)\right) = Q. \tag{4.13}$$

This equation involves the following values: the volumetric water content θ; the water saturation $S = \frac{\theta}{\varphi}$; the pressure head $\psi = \frac{p_w}{\rho g}$, p_w being pressure in the water phase; and the relative permeability $k_r(\theta)$. $\mathbb{K}$ here denotes hydraulic conductivity tensor at full saturation. The second term stands for the media and solution compressibility and is important mainly in saturated conditions.

To complete the flow model (4.13), one has to supply equations of state, relating θ and ψ and defining the function $k_r(\theta)$. There is a wide variety of these equations of state: Brooks and Corey [42], van Genuchten–Mualem [146], and many others. The van Genuchten–Mualem model proposes the following equations of state:

$$\theta(\psi) = \begin{cases} \theta_r + \frac{\theta_s - \theta_r}{[1+|\alpha_{vg}\psi|^n]^m} \text{ if } \psi < 0 \\ \theta_s \text{ if } \psi \geq 0, \end{cases} \tag{4.14}$$

$$k_r(\theta) = S^{\frac{1}{2}} \left[1 - (1 - S^{\frac{1}{m}})^m \right]^2, \tag{4.15}$$

where $m = 1 - 1/n$ and $n > 1$. Coefficients n, α_{vg} are the model parameters stemming from the media hydraulic properties. The aforementioned models of flow in unsaturated conditions are implemented in FEFLOW [55], HYDRUS [145], GeRa [91], and a couple of other hydrogeological codes. They are quite demanding in terms of media parameters which need to be measured in order to define the constitutive relations. Often these parameters are not available, and in this case it is used to apply models which are less physically justified but built upon the assumption that the flow mainly occurs in the saturated zone and tends to zero with the decrease of saturation. One approach here is the moving mesh technique proposed by Diersch [53]. It is quite demanding in implementation and has difficulties in the solution of problems having saturated areas above the free surface (see [55]). Another approach is the application of pseudo-unsaturated model on fixed grids, which is implemented, for example, in FEFLOW (see par. 9.5.4 in [55]) and MODFLOW [78] numerical codes. The model in MODFLOW does not use Richards equation; instead, it introduces variable hydraulic conductivity and variable storage term in the basic groundwater flow equation and uses techniques like inactivation of dry cells. It should be noted here that these models are not purely mathematical, but they are formulated relying on the grid already existing in the computational domain.

We will formulate the model which is implemented in GeRa code [25]. A similar model is used in FEFLOW, where it is formulated for vertex-collocated primary unknowns and with a different dependence of the moisture content on the piezometric head. We reformulate (4.13) in terms of piezometric head $\mathrm{h} = \psi + z$, where z is the vertical coordinate:

$$\frac{\partial \theta}{\partial t} + Ss_0 \frac{\partial \mathrm{h}}{\partial t} - \operatorname{div}\left(k_r(\theta)\mathbb{K}\nabla \mathrm{h}\right) = Q. \tag{4.16}$$

Piezometric head will be further used as the primary variable. A piecewise linear function is proposed for the retention curve, in which the moisture content is assumed to depend on the piezometric head rather than the pressure head as in (4.14). There are significant differences in the derivation of the model (4.16). First, expressions depend on spatial discretization, unlike in unsaturated flow models, where they depend only on the medium properties. In fact, the functions may vary from cell to cell. Second, the assumed hydrophysical properties do not have any experimental base and are introduced only as a method of approximation under the following assumptions:

- The relative permeability in unsaturated zone is small compared to saturated zone (degree of smallness is user-specified).
- The relative permeability for a cell varies linearly from minimal value (water content is close to zero) to maximal value (full saturation) with the growth of the groundwater level.

The constitutive relations used in the model are very similar to those proposed by Diersch in [55]. Here the model is formulated in terms of cell-collocated unknowns (piezometric head) for finite volume method, while for FEFLOW [55] the model is written in terms of vertex-centered unknowns (pressure head) for P_1 finite element method. Also, there is a difference in the water content calculation in unsaturated conditions (linear here versus constant in [55]). The dependence of water content θ_ω in cell ω on piezometric head h_ω in this cell is defined as follows:

$$\theta_\omega(\mathrm{h}_\omega) = \begin{cases} \varphi \text{ if } \mathrm{h}_\omega > \mathrm{h}_{\omega,\max}, \\ \varphi \frac{\mathrm{h}_\omega - \mathrm{h}_{\omega,\min}}{\mathrm{h}_{\omega,\max} - \mathrm{h}_{\omega,\min}} \text{ if } \mathrm{h}_{\omega,r} < \mathrm{h}_\omega \le \mathrm{h}_{\omega,\max}, \\ \varphi(\alpha_\varphi - \alpha_\theta(\mathrm{h}_{\omega,r} - \mathrm{h}_\omega)) \text{ if } \mathrm{h}_\omega \le \mathrm{h}_{\omega,r}, \end{cases} \tag{4.17}$$

where $h_{\omega,\max}$ and $h_{\omega,\min}$ are the maximal and minimal vertical coordinates of the cell ω nodes and $h_{\omega,r}$ is such that water content calculated by the second linear part in (4.17) is equal to $\varphi\alpha_\varphi$, namely,

$$\mathrm{h}_{\omega,r} = \mathrm{h}_{\omega,\min} + \alpha_\varphi(\mathrm{h}_{\omega,\max} - \mathrm{h}_{\omega,\min}). \tag{4.18}$$

The function $\theta_\omega(\mathrm{h}_\omega)$ therefore has three linear parts:

- It is equal to the porosity when the cell is fully saturated.
- In partially saturated conditions, it approximates the saturated volume of the cell.
- After it becomes less than $\varphi \cdot \alpha_\varphi$, it decreases slowly. One should choose small enough α_θ to guarantee that the water content value will not be negative.

The relative permeability is assumed to be equal to the saturation:

$$k_r(\mathrm{h}_\omega) = S(\mathrm{h}_\omega) = \frac{\theta_\omega(\mathrm{h}_\omega)}{\varphi}. \tag{4.19}$$

As a result, the relative permeability for the cell in saturated conditions is equal to 1. When drying the cell, the relative permeability decreases linearly while $\mathrm{h} \ge \mathrm{h}_r$. This is an approximation of the water saturation as the part of the cell volume filled by water. When $\mathrm{h} < \mathrm{h}_r$, the function $k_r(\mathrm{h}_\omega)$ is also linear.

Thus, the proposed unconfined flow model has two parameters α_φ and α_θ, which should be small enough. When these values tend to zero, water flow through the dry cells tends to zero as well.

The transport equation now has to take into account the variable saturation. This affects time derivative, diffusion, and decay terms in (4.2):

$$\frac{\partial\,(\theta + \rho_b K_d)\,c}{\partial t} - \operatorname{div} \mathbb{D}(S)\nabla c + \operatorname{div}\,(\mathbf{q}c) + \lambda\,(\theta + \rho_b K_d)\,c = f. \tag{4.20}$$

We assume that molecular diffusion is proportional to saturation, say $\mathbb{D}_m(S) = S d_m \mathbb{I}$ is substituted into formula (4.5) for $\mathbb{D}$.

4.2.2 Numerical Solution Aspects

Due to stability requirements, the flow equations are normally solved using implicit schemes in time. After discretization in space, we obtain a system of nonlinear equations with respect to grid unknowns. It may be solved by Picard or Newton iterations. The flow equation (4.13) has been written in the so-called mixed form ($\theta - h$ form), say in terms of both water content θ and pressure head ψ. It can be rewritten in the equivalent so-called pressure form (h-form) by replacing the time derivative term

$$\frac{\partial \theta}{\partial t} = \frac{\partial \theta}{\partial \psi}\frac{\partial \psi}{\partial t}. \tag{4.21}$$

The derivative $\dfrac{\partial \theta}{\partial \psi}$ is called the specific moisture capacity. The solution of the flow equation in the pressure form is prone to mass balance violation especially in zones of strong nonlinearity of the retention curve, for example, when modeling irrigation of dry soils. To overcome this nonconservativity, Celia et al. [44] proposed a modification to Picard iteration algorithm allowing to solve the equation in the mixed form having only pressure unknowns on each Picard iteration. This is achieved by the following θ time derivative discretization:

$$\frac{\theta^{n+1,k+1} - \theta^{n}}{\Delta t} \approx \frac{\theta^{n+1,k} - \theta^{n}}{\Delta t} + \frac{\partial \theta}{\partial \psi}(\psi^{n+1,k})\frac{\psi^{n+1,k+1} - \psi^{n+1,k}}{\Delta t}, \tag{4.22}$$

in which indices n and k stay for the time step and Picard iteration number, respectively. The first term of this expression is known from the previous iteration. The system at $(k+1)$th iteration then takes the following form:

$$\mathbb{A}(\psi^{n+1,k})\psi^{n+1,k+1} = f(\psi^{n+1,k}), \tag{4.23}$$

where $\mathbb{A}(\psi^{n+1,k})$ is the matrix of coefficients evaluated on the previous iteration, $f(\psi^{n+1,k})$ is the right-hand side collecting all known terms. The initial guess may be chosen as the solution on the previous time step, say $\psi^{n+1,0} = \psi^{n}$. Stopping criterion for Picard iterations may be based on relative or absolute residual reduction for the system (4.23) or on the reduction of the difference between solutions (ψ^{n+1} and/or θ^{n+1}) on two successive iterations. The problem with the Picard method is that it often requires small time steps to converge, especially for strongly heterogeneous soils or dry conditions. A smaller time step provides contraction property of the corresponding mapping. In order to achieve convergence, in hydrogeological codes adaptive time-stepping algorithm is usually implemented (for example, in FEFLOW, HYDRUS, GeRa).

Application of the Newton method for the nonlinear systems solution is aimed not only to accelerate convergence but also to avoid the necessity of time step refinement only to make the method convergent. It is successfully implemented for modeling groundwater flow in unconfined conditions in MODFLOW-NWT [116] and FEFLOW [55]. Newton method is implemented in GeRa as well for the linear TPFA.

The nonlinear system for nth time step may be rewritten in the form

$$\mathcal{R}(\psi^{n+1}) \equiv \mathbb{A}(\psi^{n+1})\psi^{n+1} - f(\psi^{n+1}) = 0, \tag{4.24}$$

and a linear system with the Jacobian matrix $\mathcal{J} : J_{ij} = \frac{\partial \mathcal{R}_{\omega_i}}{\partial \psi^{n+1}_{\omega j}}$ is formed and solved at $(k+1)$th iteration:

$$\mathcal{J}(\psi^{n+1,k})(\psi^{n+1,k+1} - \psi^{n+1,k}) = -\mathcal{R}(\psi^{n+1,k}). \tag{4.25}$$

The stopping criteria here are analogous to those of Picard iterations and are aimed at residual reduction.

Our practice shows that the Picard method needs to use relaxation (consecutive solution of unsteady flow problems till stationary solution is achieved) to solve stationary flow problems while the Newton method often is able to solve them without any extra techniques. In order to further improve convergence of the Newton method, Forsyth et al. [67] proposed primary variable switching technique. It offers to solve the flow equations in terms of ψ in cells with high saturation and in terms of θ in sufficiently unsaturated cells. This technique was thoroughly investigated by Diersch and Perrochet [54] within different time-stepping schemes and showed a significant speedup and robustness with respect to solution in terms of ψ in case of very dry soil initial conditions.

The Picard and Newton methods may be applied for the solution of flow in unconfined conditions problems of the form (4.16), (4.17). Picard iterations have shown to strongly suffer from grid refinement in the z-direction because of the increase in the steepness of the function $\theta(\mathrm{h})$ defined by (4.17). Severe time step refinement is necessary to ensure the mapping contraction property. In these cases, Newton method allows to reduce calculation time by several orders of magnitude being able to make large time steps [25]. To further accelerate convergence of both Picard and Newton iterations, a correction method is proposed on each iteration in [25]. After iteration of a nonlinear solver is completed and new head values are calculated, some cells may change their saturation state (see three cases in (4.17)). We focus on two types of state changes. The first one is the shift from dry state to partially saturated state which happens in case if the new piezometric head value h_ω in the cell ω is bigger than $\mathrm{h}_{\omega,r}$ and the previous head value $\mathrm{h}_{\omega,prev}$ is less than $\mathrm{h}_{\omega,r}$. The second one is the shift from completely to partially saturated state which occurs if the new head value h_ω is less than $\mathrm{h}_{\omega,max}$ and the previous head value $\mathrm{h}_{\omega,prev}$ is bigger than $\mathrm{h}_{\omega,max}$. In both cases, the new head value may be corrected, because it was calculated using specific moisture capacity $\frac{\partial \theta_\omega}{\partial \mathrm{h}}$ in the formula (4.22) corresponding to another linear part of function (4.17). The idea of the proposed correction is to calculate the new value of h_ω providing a proper (conservative) change in water content in the cell ω taking into account the dependence $\theta_\omega\,(\mathrm{h}_\omega)$ defined by (4.17).

In case of shift from dry to partially saturated state, the corrected value is defined as

$$\mathrm{h}_{\omega,new} = \min\left\{\mathrm{h}_{\omega,\max}, \mathrm{h}_{\omega,r} + (\mathrm{h}_{\omega} - \mathrm{h}_{\omega,r}) \cdot \alpha_{\theta}(\mathrm{h}_{\omega,\max} - \mathrm{h}_{\omega,\min})\right\}. \tag{4.26}$$

In case of shift from completely to partially saturated state, the corrected value is defined as

$$\mathrm{h}_{\omega,new} = \max\left\{\mathrm{h}_{\omega,r}, \mathrm{h}_{\omega,\max} + (\mathrm{h}_{\omega} - \mathrm{h}_{\omega,\max}) \cdot \frac{s_0(\mathrm{h}_{\omega,\max} - \mathrm{h}_{\omega,\min})}{\varphi}\right\}. \tag{4.27}$$

As shown in [25], this correction allows to accelerate both Picard and Newton solutions of the flow in unconfined conditions problems up to two times.

4.2.3 Numerical Examples

4.2.3.1 Dam Seepage Problem

The two-dimensional dam seepage problem can be used to verify the unconfined flow numerical model. The dam is represented as a square 10 m × 10 m. Constant piezometric head h = 10 m is prescribed on the left boundary and h = 2 m on the right boundary below $z = 2$ m. Above this level, a seepage face boundary condition is prescribed on the right boundary. Top and bottom boundaries are impermeable. The dam is homogeneous and has isotropic hydraulic conductivity $K_s = 10^{-5}$ m/s. The flow through it can be calculated using the Dupuit formula [46]:

$$Q = Kb\frac{H_1^2 - H_2^2}{2l}, \tag{4.28}$$

b is the dam width, H_1 and H_2 are the water levels on upstream and downstream sides, respectively, and l is the dam width (see Fig. 4.1). For the aforementioned problem parameters, the flux shall be 4.8E-5 m^3/s assuming $b = 1$ m. According to the analytical steady-state solution which can be found in [124], the water level H_0 should be at the height of approximately 4 m (the value was obtained from the plot in [124]) at the right boundary. The aforementioned model of flow in unconfined conditions is used with two sets of parameters α_{φ}, α_{θ} in order to assess their influence on the results. From Tables 4.1, 4.2, one can see that H_0 converges to the analytical value with mesh refinement. For the first set of parameters $\alpha_{\varphi} = 10^{-2}$, $\alpha_{\theta} = 10^{-3}$, we observe stagnation of convergence after the space step is dropped below 0.2 m. But when these parameters are chosen larger, flux convergence stagnates for cell size 0.1 m and less. This is likely to be the result of error produced by the mathematical model itself, while the approximation error is already below it. To prove this statement, a calculation with an order smaller parameters $\alpha_{\varphi} = 10^{-3}$, $\alpha_{\theta} = 10^{-4}$ is performed (see Table 4.2). With these parameters, the model error is smaller than

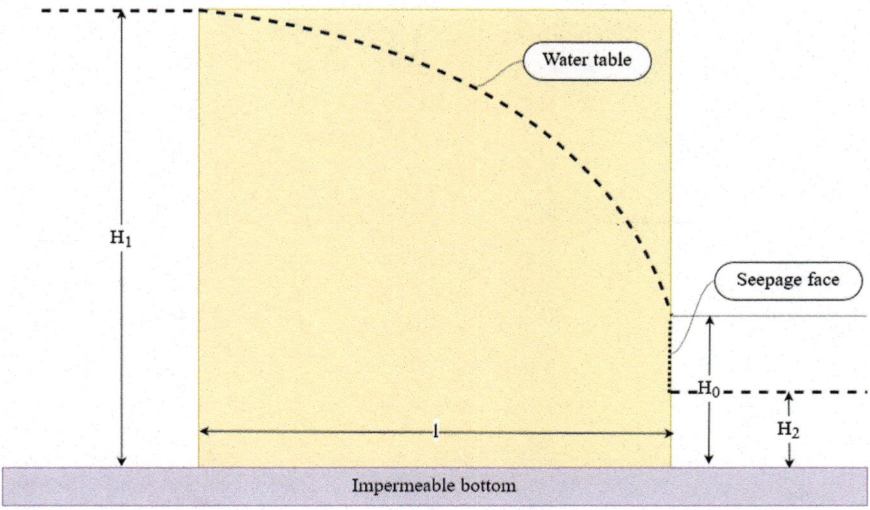

Fig. 4.1 Geometry of the dam seepage problem

Table 4.1 Dam seepage problem calculation results for $\alpha_\varphi = 10^{-2}$, $\alpha_\theta = 10^{-3}$

Cell size	H_0	Q	Calc. time (s)	No. of Newton iterations
0.4	4.40	4.77815E-5	<1	55
0.2	4.20	4.79575E-5	4	77
0.1	4.10	4.80213E-5	22	98
0.05	4.05	4.80454E-5	110	114
0.025	4.00	4.80546E-5	734	131

Table 4.2 Dam seepage problem calculation results for $\alpha_\varphi = 10^{-3}$, $\alpha_\theta = 10^{-4}$

Cell size	H_0	Q	Calc. time (s)	No. of Newton iterations
0.4	4.40	4.77398E-5	<1	53
0.2	4.20	4.79088E-5	5	99
0.1	4.10	4.79697E-5	40	189
0.05	4.05	4.79906E-5	326	360
0.025	4.00	4.79995E-5	3297	765

the approximation error and flux convergence to the analytical value is observed for the whole sequence of meshes. Calculation times obtained on the same personal computer (Intel Core i7 based) and the number of Newton iterations are also given in the tables. Calculated location of the water table for the dam see page problem is shown in Fig. 4.2.

Fig. 4.2 Calculated zone of full saturation in the dam

4.2.3.2 Capillary Barrier Test

The capillary barrier problem is often used for testing the correctness and robustness of unsaturated flow models. Modeling results are normally compared to Ross' formula for the diversion length [126]. The test was first introduced by Oldenburg and Pruess in [117] for testing TOUGH2 software, but their solution showed poor agreement with analytics. Later, Webb [151] extended Ross' formula for van Genuchten media hydraulic properties and showed reasonable numerical results using TOUGH2 with upstream weighting for permeability. The problem was also used by Diersch [55] to verify FEFLOW software as well.

Capillary barriers appear in unsaturated conditions when a fine-grained layer overlays a coarse-grained one with a tilted interface between them. In this case, water infiltrating from the top of the fine-grained layer is diverted by the interface and flows down within the top layer due to capillary forces. As saturation grows downdip the surface, moisture starts to leak through the barrier reaching at some point the maximum leakage which is equal to infiltration rate. This point is called capillary barrier diversion length and is measured as the distance down the slope of the surface, at which the leakage/infiltration ratio becomes 1.

The problem setting is shown in Fig. 4.3, and the computational domain is the same as in [55, 151]. The domain contains two 5% tilted layers with thickness 0.5 m each; a fine-grained layer is above a coarse-grained one. The length of the domain is 100 m. We assume that this length is enough to make the effects of boundary conditions on the right side negligible for the diversion length which is estimated at 32.6 m in [55]. Hydraulic parameters of the media are provided in Table 4.3, and the function

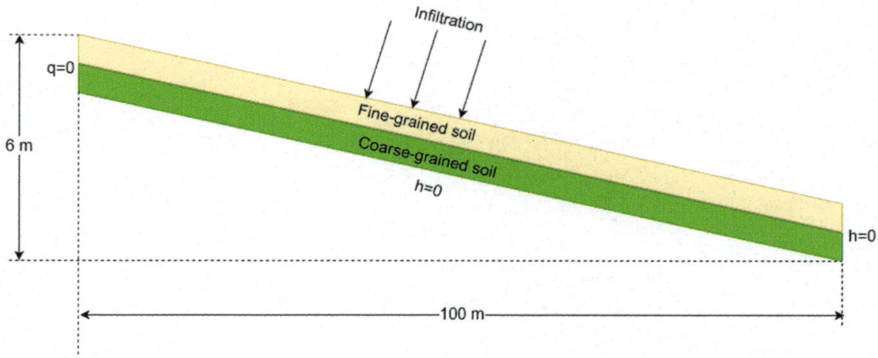

Fig. 4.3 Piezometric head distribution for the capillary barrier test

Table 4.3 Media hydraulic parameters

Parameter	Fine-grained soil	Coarse-grained soil
θ_S	0.39	0.42
θ_r	0.154	0.012
K_s, cm/s	0.021	10
n	5.74	2.19
α, cm^{-1}	0.039	4.9

$K(\psi)$ plot is shown in Fig. 4.4. On the left vertical boundary, zero flux condition is prescribed. On the bottom and the right boundaries, Dirichlet boundary condition $h = \psi + z = 0\,\text{m}$ is set, which provides water table location at the height of the bottom right corner of the domain and hydrostatic conditions along the bottom edge. Infiltration boundary condition is prescribed on the top boundary $q = 0.48\,\text{cm/day}$ (flux in the vertical direction).

For simulation in GeRa, a hexahedral grid with 1600 cells was generated. The computational domain was divided into four grid subdomains: the top and bottom layers 0.45 m thick and two middle layers 0.05 m thick. The mesh edges are aligned with the domain boundaries. Every grid subdomain contains an equal number of grid layers, namely, 8. Thus, the grid is uniform in x-direction with space step 2 m and refined to the materials' interface in z-direction. In the middle layers, the vertical space step is 0.00625 m, while in the top and bottom layers it is 0.05625 m.

The computational domain is 100 m long (in the x-direction). The calculated piezometric head in the domain is shown in Fig. 4.5. In Fig. 4.6, one can see the calculated leakage/infiltration ratio which provides good qualitative and quantitative agreement with Ross' formula and numerical results of Webb [151] and Diersch [55]. It should be noted that in the vicinity of the right boundary the present solution is different from these mentioned above. Close to the right boundary, the flux through the fine-coarse interface reasonably decreases due to the influence of Dirichlet-type boundary condition (h = 0). This effect is also proved by the satu-

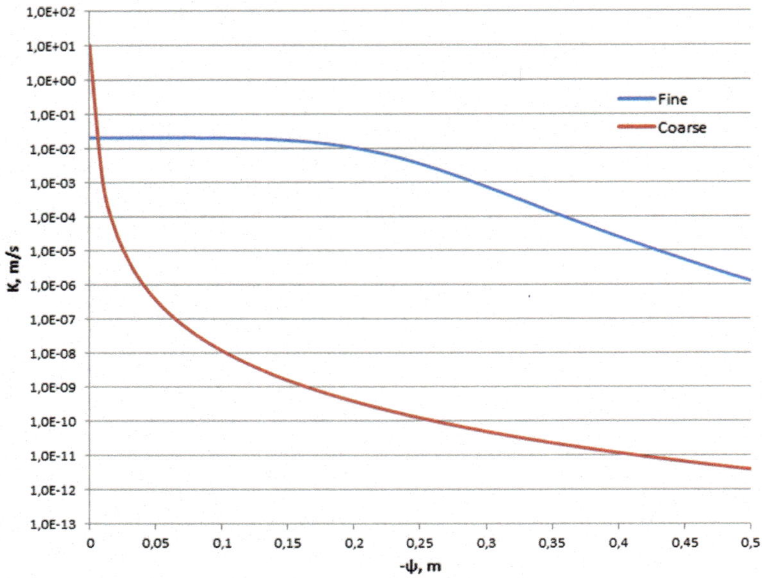

Fig. 4.4 Hydraulic conductivity dependence on pressure head

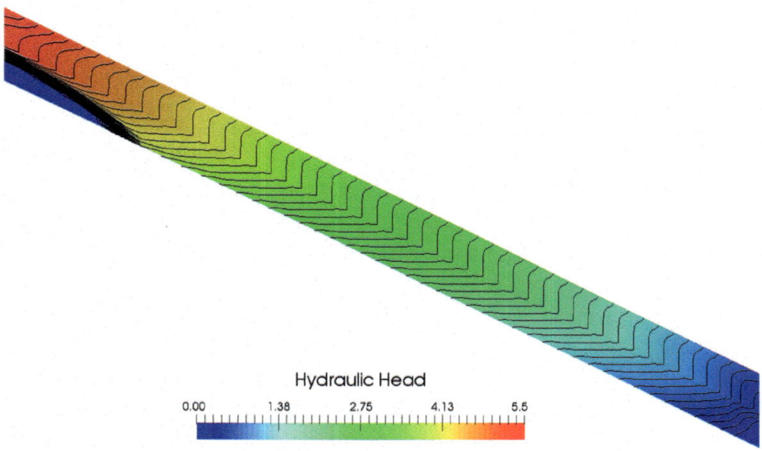

Fig. 4.5 Calculated piezometric head distribution for the capillary barrier test

ration plot in [55]. Comparing the use of three different cell-centered finite volume schemes, namely, linear TPFA, linear MPFA O-scheme, and monotone scheme with nonlinear TPFA for the capillary barrier problem, we observe almost identical results for leakage/infiltration ratio. This also holds for central and upwind relative permeability approximation, showing negligible difference in the results.

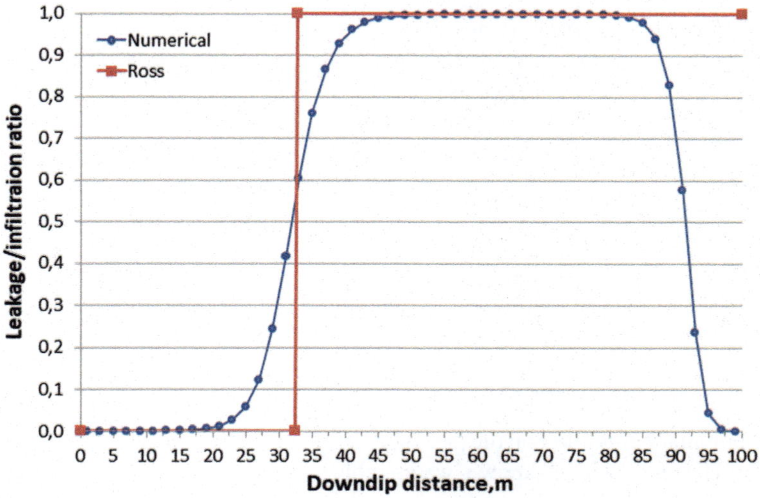

Fig. 4.6 Leakage/infiltration ratio calculated using the MPFA-O-scheme

4.2.3.3 Surface Landfill Site

Surface landfills are one of the major ecological problems which need modeling flow and transport in unsaturated conditions. Here, an example of a real landfill is studied with the aim to demonstrate the abilities to solve unsaturated flow and transport problems with highly anisotropic and in homogeneous media parameters. In Fig. 4.7, a map of the area with the model boundaries and location of the landfill is shown. In Fig. 4.8, one can see the relief of the site.

The geological model includes seven layers (see Fig. 4.9) composed of eight materials with different properties. The top layer (I) corresponds to technogenic materials (debris) and contains solely the landfill. Layers II, IV, and VI are aquifers, while layers III, V, and VII are aquitards. Material 1 is the debris and occupies the body of the landfill which is located in the two top layers within the landfill boundaries. The other layers are composed of materials with the corresponding numbers. The last eighth material occupies a part of the fourth layer.

Material properties are given in Table 4.4. The first material features anisotropic hydraulic conductivity defined by horizontal coefficient K_{x-y} and vertical K_z, while the other materials are isotropic. The problem features highly heterogeneous parameters, namely, hydraulic conductivity varies by six orders of magnitude and parameters of the van Genuchten retention curve vary from those specific for clays (material 7) to the ones specific for sands (materials 4, 6, and 8).

Table 4.4 Landfill site hydraulic parameters

Material #	K_{x-y} (m/day)	K_z (m/day)	θ_S	θ_r	α (m^{-1})	n
1	0.5	0.001	0.40	0.1	6.0	1.50
2	5	5	0.41	0.065	7.5	1.89
3	0.005	0.005	0.43	0.078	3.6	1.56
4	1.3	1.3	0.43	0.045	14.5	2.68
5	0.006	0.006	0.43	0.078	3.6	1.56
6	3	3	0.43	0.045	14.5	2.68
7	2.5E-6	2.5E-6	0.38	0.068	0.8	1.09
8	4	4	0.43	0.045	14.5	2.68

The boundary condition on the part of the planar boundary marked blue in Fig. 4.7 is set as Dirichlet with a known water table height h = 135 m. This part of the boundary corresponds to groundwater level isoline and is located upstream. The other planar boundaries are considered as impermeable. Rainfall recharge boundary conditions are set on top of the domain with higher recharge in the landfill area, namely, 0.0002 m/day and 0.0005 m/day. The bottom boundary is the contact with an underlying aquifer, and constant piezometric head h = 101 m is assumed on it. Thus, in the model, groundwater may discharge into two existing rivers or into the deep aquifer below the computational domain.

Stationary flow in unsaturated conditions is simulated. For transport modeling, a point source of non-sorbing contaminant with constant intensity is placed approximately in the middle of the landfill. The calculated saturation in the computational domain is shown in Fig. 4.10 along with the grid. The grid contains 338430 cells, namely, 325224 triangular prisms, 6396 tetrahedra, and 6810 pyramids. Tetrahedra and pyramids appear due to layers' pinch-outs. Saturation is high within the body of the landfill due to high infiltration rate on its top and in the vicinity of the rivers which act as discharge areas. The contaminant plume for the time 200 years is shown in Fig. 4.11. Pollution moves mainly in the vertical direction within the landfill body reaching layer 6 due to lack of aquitards' integrity. Within layer 6, it is driven to the discharge area located at the river. Due to the complicated structure of the flow, the plume splits into two parts.

4.3 Reactive Transport Modeling

4.3.1 Mathematical Models

In this section, we consider reactive transport modeling taking into account homogeneous reactions of aqueous species and heterogeneous reactions of sorption. Comprehensive description of reactive transport models including equilibrium and kinet-

Fig. 4.7 Computational domain footprint and the landfill location on the map (*source:* https://www.openstreetmap.org)

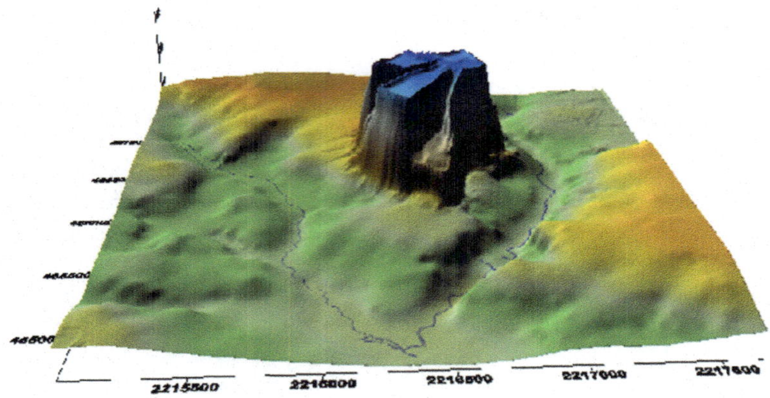

Fig. 4.8 Relief of the landfill site

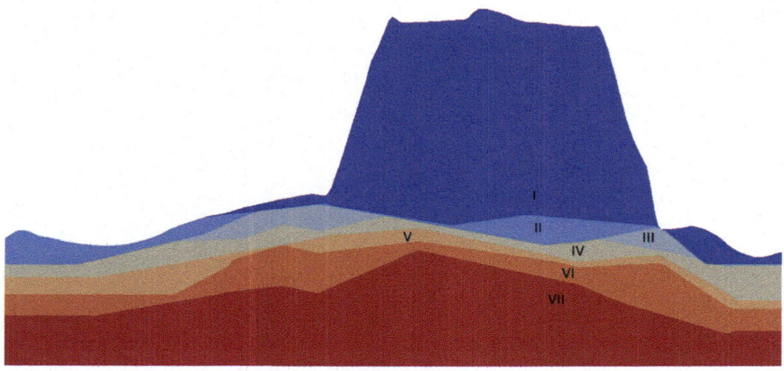

Fig. 4.9 Lateral geological profile passing through the landfill

ically controlled chemical reactions can be found in [31] or [137]. In the following, we will neglect kinetically controlled reactions assuming that the specific timescale of reactions is much smaller than the one for transport processes. Let us consider a system containing $N_{tot} = N_{aq} + N_{sorb}$ chemical species with N_{aq} aqueous and N_{sorb} sorbed species. Chemical equilibrium condition results in a system of $N_{sec} + N_{sorb}$ algebraic equations expressing mass action law which allows to reduce the number of unknowns in the transport equation, N_{sec} being the number of so-called secondary species in the system. Following Saaltink et al. [129] one can choose a set of $N_{pr} = N_{tot} - N_{sec} - N_{sorb}$ so-called primary species or components using these mass action law equations. Then, disregarding precipitation–dissolution and source terms and assuming a unique molecular diffusion coefficient for all species for the sake of simplicity, one can write the following system of equations which defines the reactive transport model:

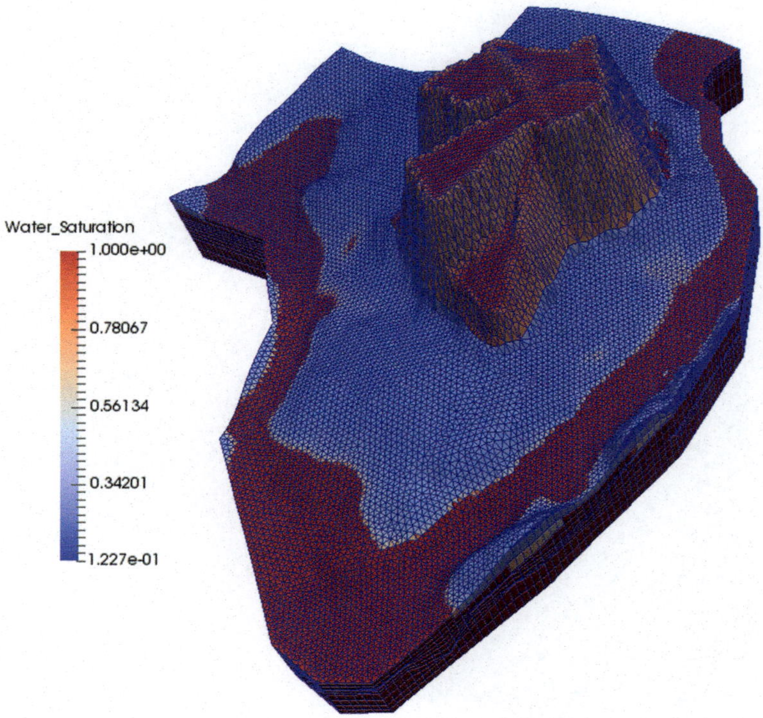

Fig. 4.10 Saturation distribution and the computational grid (scaled 10 times in z-direction)

$$\varphi\left(\frac{\partial c_j}{\partial t}+\frac{\partial s_j}{\partial t}\right)=\operatorname{div}\mathbb{D}\nabla c_j-\operatorname{div}\left(\mathbf{q}c_j\right),\quad j=1,\ldots,N_{pr}. \tag{4.29}$$

Equation (4.29) is written in terms of total primary species concentration, i.e. c_j and s_j are component concentrations in aqueous and sorbed forms, respectively. They are expressed in terms of primary, secondary, and sorbed species by the formula

$$c_j=\tilde{c}_j+\sum_{i=N_{pr}}^{N_{pr}+N_{sec}}\nu_{ij}\tilde{c}_i,\ j=1,\ldots,N_{pr}; \tag{4.30}$$

$$s_j=\sum_{i=N_{pr}+N_{sec}+1}^{N_{tot}}\nu_{ij}\tilde{c}_i,\ j=1,\ldots,N_{pr}. \tag{4.31}$$

Here, $\tilde{c}_i, i=1,\ldots,N_{tot}$ are concentrations of N_{aq} aqueous species and N_{sorb} adsorbed species, ν_{ij} are the stoichiometric coefficients defined by the equilibrium reactions between primary and secondary or primary and sorbed species in the form

Fig. 4.11 Contamination plume at 200 years (scaled 10 times in z-direction)

$$A_i = \sum_{j=1}^{N_{pr}} \nu_{ij} A_j, \tag{4.32}$$

where A_i and A_j are chemical formulas of the secondary (or sorbed) and primary species, respectively. Finally, the secondary and sorbed species' concentrations are bound to the primary species' concentrations by the mass action law equations

$$K_i a_i = \prod_{j=1}^{N_{pr}} \left(a_j\right)^{\nu_{ij}}, \quad i = N_{pr} + 1, \ldots, N_{tot}, \tag{4.33}$$

where a_i is the activity of species i and K_i is the equilibrium constant for the reaction. It is assumed that the activities can be expressed using the formula

$$a_i = \gamma_i \tilde{c}_i, \quad i = 1, \ldots, N_{tot}, \tag{4.34}$$

with γ_i being the activity coefficient for species i.

To the best of our knowledge, the model (4.29)–(4.31), (4.33), (4.34) is used mainly for near-field modeling or to substantiate the distribution coefficient K_d. The reason is, first of all, high computational load caused by chemical equilibrium calculations, which demands high-performance computing especially in case when uncertainty quantification is needed. Little knowledge on solutes' and rock's chemical composition on a large space scale and high degree of uncertainty in these parameters also make these models difficult to apply in the far-field. In the latter case, simple models with sorption governed by isotherms (4.2), (4.4) are preferred. At the same time, in order to account geochemistry at affordable costs, a variable K_d (also known as "Smart K_d") approach can be used [34, 139]. Its main idea is to determine a set of environmental parameters (e.g., some chemical species concentration, temperature, or mineralogy) most affecting sorption of contaminants. For each sorbed species, a multidimensional table is built containing K_d values calculated for various combinations of these key parameters. In the transport calculations, one has to track pollutants along with the species needed to determine the key parameters. For example, in order to control pH and redox potentials, one needs to calculate transport of hydrogen, oxygen, and charge balance. Normally, radionuclides are present in trace concentrations, thus not affecting geochemical conditions significantly. Parameter K_d used in transport calculations is obtained using interpolation of table values with known key parameters.

4.3.2 Numerical Solution Aspects

Concerning the discretization in time, there are three major approaches to the numerical solution of reactive transport problems [137]. These are direct substitution approach (DSA), sequential non-iterative approach (SNIA), and sequential iterative approach (SIA).

DSA results in the substitution of algebraic expressions of total aqueous and sorbed primary species concentrations $c_j, s_j, j = 1, \ldots, N_{pr}$ in terms of aqueous species concentrations $\tilde{c}_j, j = 1, \ldots, N_{pr}$ using relations (4.30), (4.31) and mass action law equations (4.33). It is convenient to express explicitly the mass action law in terms of concentrations using (4.33), (4.34) in the form

$$\tilde{c}_i = K_i^{-1} \gamma_i^{-1} \prod_{j=1}^{N_{pr}} \left(\gamma_j \tilde{c}_j\right)^{\nu_{ij}}, \quad i = N_{pr} + 1, \ldots, N_{tot}, \tag{4.35}$$

which allows us to eliminate easily secondary and sorbed species concentrations. The result of substitution is a global system of N_{pr} nonlinear partial differential equations which is to be solved for primary species concentrations $\tilde{c}_{pr} = \left(\tilde{c}_1, \ldots, \tilde{c}_{N_{pr}}\right)$:

$$\varphi\left(\frac{\partial c_j\left(\tilde{c}_{pr}\right)}{\partial t}+\frac{\partial s_j\left(\tilde{c}_{pr}\right)}{\partial t}\right)=\operatorname{div}\mathbb{D}\nabla c_j\left(\tilde{c}_{pr}\right)-\operatorname{div}\left(\mathbf{q}c_j\left(\tilde{c}_{pr}\right)\right), \quad j=1,\ldots,N_{pr}. \tag{4.36}$$

After its discretization in time (implicitly) and space, one obtains a system of nonlinear algebraic equations which is usually solved by the Newton–Raphson method.

SIA allows for the solution of implicit backward Euler discretization in time of the reactive transport problem using the fixed-point iteration. On each iteration, first, the transport problem for total aqueous primary species concentration is solved evaluating sorbed concentration on the previous iteration. Second, chemical equilibrium is restored using the total aqueous concentrations from the first step. Iterations are repeated until the norm of the difference between two successive solutions decreases below some specified precision. Denoting by k the iteration count and by n the time step index, we can write these substeps in the following way:

$$1.\quad \varphi\frac{c_j^{n+1,*}-c_j^n}{\Delta t}-\operatorname{div}\mathbb{D}\nabla c_j^{n+1,*}+\operatorname{div}\left(\mathbf{q}c_j^{n+1,*}\right)= =-\varphi\frac{s_j^{n+1,k}-s_j^n}{\Delta t},\quad j=1,\ldots,N_{pr}; \tag{4.37}$$

$$2a.\quad \tilde{c}_j^{n+1,k+1}+\sum_{i=N_{pr}}^{N_{pr}+N_{sec}}\nu_{ij}K_i^{-1}\gamma_i^{-1}\prod_{l=1}^{N_{pr}}\left(\gamma_l\tilde{c}_l^{n+1,k+1}\right)^{\nu_{il}}+ +\sum_{i=N_{pr}+N_{sec}+1}^{N_{tot}}\nu_{ij}K_i^{-1}\gamma_i^{-1}\prod_{l=1}^{N_{pr}}\left(\gamma_l\tilde{c}_l^{n+1,k+1}\right)^{\nu_{il}}= =c_j^{n+1,*}+s_j^{n+1,k},\quad j=1,\ldots,N_{pr} \tag{4.38}$$

$$2b.\quad c_j^{n+1,k+1}=\tilde{c}_j^{n+1,k+1}+\sum_{i=N_{pr}}^{N_{pr}+N_{sec}}\nu_{ij}\tilde{c}_i^{n+1,k+1},\quad j=1,\ldots,N_{pr};$$

$$s_j^{n+1,k+1}=\sum_{i=N_{pr}+N_{sec}+1}^{N_{tot}}\nu_{ij}\tilde{c}_i^{n+1,k+1},\quad j=1,\ldots,N_{pr}. \tag{4.39}$$

In (4.37)–(4.39), the unknowns are put on the left-hand side, while the known values are put on the right-hand side. On the first step, an intermediate transport solution $c^{n+1,*}$ is found. The chemical step 2 contains two substeps. Given the total concentration of aqueous primary species on the current iteration $c^{n+1,k+1}$ and the total sorbed primary species concentration on the previous iteration $s_j^{n+1,k}$, the vector of aqueous primary species concentrations $\tilde{c}_{pr}^{n+1,k+1}$ is calculated from the chemical equilibrium conditions on substep 2a. Then on substep 2b, finally the total aqueous and sorbed primary species concentrations on iteration $k+1$ are obtained. If necessary, the concentrations of secondary and sorbed species may be calculated using (4.35) and substituting $\tilde{c}_{pr}^{n+1,k+1}$.

SNIA is the simplest approach to temporal discretization of reactive transport equations implemented in a vast majority of modeling codes, while only a few of them feature full implicit schemes (DSA and SIA) [137]. It is an operator-splitting scheme in which the transport problem and the chemical equilibrium problem are solved sequentially:

$$1.\quad \varphi\frac{c_j^{n+1,*}-c_j^n}{\Delta t} - \operatorname{div}\mathbb{D}\nabla c_j^{n+1,*} + \operatorname{div}\left(\mathbf{q}c_j^{n+1,*}\right) = 0,$$
$$j = 1,\ldots,N_{pr}; \tag{4.40}$$

$$2a.\quad \tilde{c}_j^{n+1} + \sum_{i=N_{pr}}^{N_{pr}+N_{sec}} \nu_{ij}K_i^{-1}\gamma_i^{-1}\prod_{l=1}^{N_{pr}}\left(\gamma_l\tilde{c}_l^{n+1}\right)^{\nu_{il}} +$$
$$+ \sum_{i=N_{pr}+N_{sec}+1}^{N_{tot}} \nu_{ij}K_i^{-1}\gamma_i^{-1}\prod_{l=1}^{N_{pr}}\left(\gamma_l\tilde{c}_l^{n+1}\right)^{\nu_{il}} =$$
$$= c_j^* + s_j^n, \quad j = 1,\ldots,N_{pr} \tag{4.41}$$

$$2b.\quad c_j^{n+1} = \tilde{c}_j^{n+1} + \sum_{i=N_{pr}}^{N_{pr}+N_{sec}} \nu_{ij}\tilde{c}_i^{n+1}, \quad j = 1,\ldots,N_{pr};$$

$$s_j^{n+1} = \sum_{i=N_{pr}+N_{sec}+1}^{N_{tot}} \nu_{ij}\tilde{c}_i^{n+1}, \quad j = 1,\ldots,N_{pr}. \tag{4.42}$$

Different temporal discretization schemes may be used for the transport step (4.40), including the ones explicit for advection and implicit for diffusion. Within SNIA, it is easy to couple existing flow and transport modeling codes with advanced geochemical codes. PHREEQC with its reaction modules IPhreeqc [45] and PhreeqcRM [122] specially designed for integration into transport codes is the most widely used geochemical engine. GeRa also utilizes SNIA with IPhreeqc as the geochemical engine, and implementation of models with sorption governed by isotherms with variable K_d is also accomplished via the operator-splitting scheme.

4.3.3 Numerical Examples

4.3.3.1 Iron Removal Water Treatment

This test shows the ability of the code to model advective–diffusive reactive transport in case when correct calculation of pH and pe potentials is vital. Dissolved iron is a pollutant often present in artesian water supply wells. One of possible technologies

Table 4.5 Solutions parameters

Parameter	Native solution	Injected solution
pH	4.0	7.0
pe	7.6266	13.674
Fe(2), mol/l	0.001	0
Cl, mol/l	0.002	0.01
Fe(3), mol/l	1.0E-6	0
Na, mol/l	0	0.01
O(0), mol/l	0	0.001

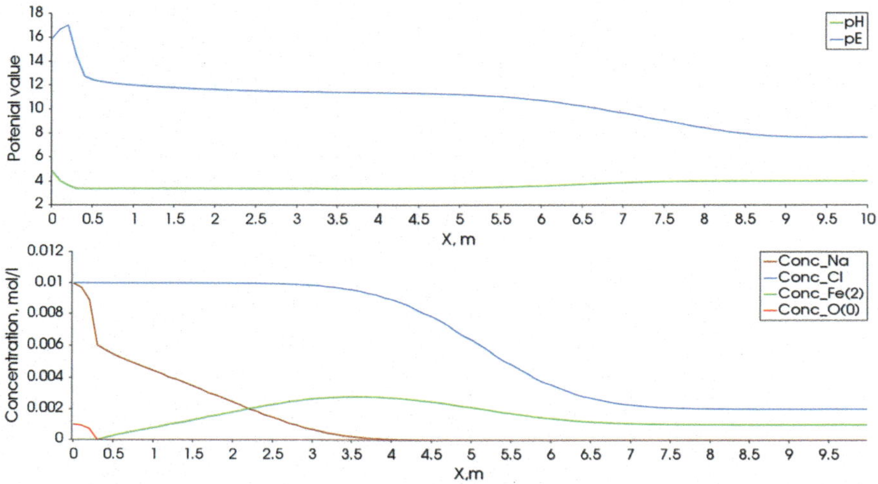

Fig. 4.12 Potentials (top) and species (bottom) distribution in the column at T = 10 days

of iron elimination is subsurface water and rock pretreatment by highly oxidized water. It allows to convert two-valent iron Fe(2) contained in dissolved and sorbed forms to three-valent form Fe(3), which precipitates within Fe_2O_3. Here, uniform flow in one-dimensional column is modeled. The column is 10 m long, porosity is 0.2, dispersivity is 0.1 m, and Darcy velocity is 0.2 m/day. One pore volume passing through the column is modeled, say the modeling period is 20 days. Highly oxidized solution containing NaCl is applied to the inflow (left) boundary of the column. The initial solution in the column is equilibrated with the media containing 0.1 mol/l of exchange phase X. Solution compositions are given in Table 4.5.

Simulation results in Figs. 4.12 and 4.13 show that Na replaces sorbed Fe(2) by the ionic exchange mechanism. Thus, one can see an increase of dissolved iron concentration above the initial value in the native solution. Dissolved oxygen O(0) is spent on oxidation of Fe(2) in the beginning of the column. Thus, an area free of both dissolved and sorbed Fe(2) is formed. Cl acts as a neutral tracer and shows the

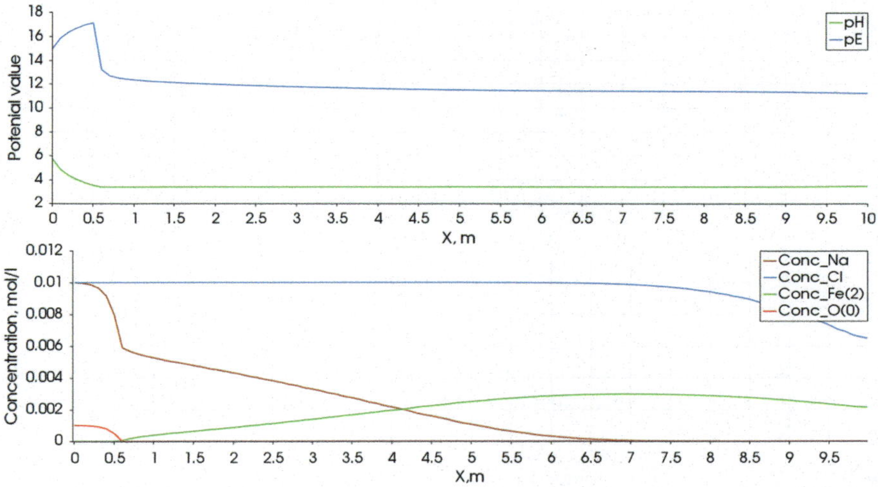

Fig. 4.13 Potentials (top) and species (bottom) distribution in the column at T = 20 days

position of the displacement front. The decrease of pH is induced by hydrolysis of generated Fe(3).

4.3.3.2 Influence of Nitrate Concentration on Sorption of Strontium

In this numerical experiment, we will focus on problems of deep well liquid radioactive waste injection sites [127]. The injected solutions contain high concentration of sodium nitrate ($NaNO_3$) up to 240 g/l, which may significantly affect the media sorption properties. It was proved experimentally that $K_d(Sr)$ decreases with the growth of $NaNO_3$ [34]. The two models introduced in Sect. 4.3.1 are compared, the first is a reactive transport model with equilibrium chemistry and the second one applies the smart-K_d approach. For the smart-K_d model, nitrate ion is chosen as the key environmental parameter.

The computational domain is a parallelepiped 200 × 200 × 10m with four injection wells close to its corners and one production well in the center. The problem may be treated as two-dimensional, as single layer of cells is used for vertical discretization. The injection wells have the rate 10 m^3/day, while the rate of the production well is 40 m^3/day. Diffusion and dispersion are neglected. The chemical parameters of the media are chosen in a way to simulate the properties of the real geological aquifer used for liquid radioactive waste injection. Its samples, being a mixture of sand and clay, were used in the sorption experiments in [34].

Composition of the native solution in the aquifer is given in Table 4.6; its pH is 7. The media contains two ionic exchange phases (X and Noneselective) equilibrated with the native solution. Porosity is 0.3. The injected solutions are identical to the

Table 4.6 Native solution parameters

Species	Concentration (mol/l)
Na	3.21E-3
Ca	4.13E-4
Cl	3.95E-4
K	2.56E-4
Mg	1.98E-4
S	5.90E-4
C	3.21E-3

native one except for the concentrations of $NaNO_3$ and Sr (radioactive component). The concentration of Sr in all injection wells is 3.3E-9 mol/l. Concentration of $NaNO_3$ is variable:

- 0.0 mol/l in the well P_1 located in the bottom left corner;
- 0.2 mol/l in the well P_2 located in the bottom right corner;
- 0.4 mol/l in the well P_3 located in the top right corner;
- 0.6 mol/l in the well P_4 located in the top left corner.

Simulation was performed for time period of 3000 days. For geochemical calculations, the chemical database *llnl.dat* provided with PHREEQC version 2.18 was augmented by a set of reactions mentioned in [34].

Results are presented in Figs. 4.14, 4.15, 4.16, and 4.17, where isolines of Sr concentration 10^{-10} mol/l and 10^{-9} mol/l and nitrate ion concentration from 0.1 mol/l to 0.6 mol/l with step 0.1 mol/l are shown. Both models demonstrate significant influence of chemical composition on the propagation of radioactive strontium, which can not be treated using models with fixed K_d. Transport of strontium is accelerated by high nitrate ion concentration. Comparing Figs. 4.14 and 4.15, one can observe faster strontium migration in the variable K_d model for $NaNO_3$ concentration above 0.2 mol/l and slower for zero $NaNO_3$ concentration in the injected solution. Taking into account the usual high level of uncertainty in the geological media and solution parameters, the difference between these models can be considered as acceptable, and variable K_d approach may be applied in order to obtain conservative (worst case) results important in the framework of a RW repository safety assessment. The corresponding table of $K_d\left(c\left(NO_3^-\right)\right)$ dependence shall be corrected in order to ensure a pessimistic propagation speed. One important outcome of this comparison is the evident retardation of nitrate ion, which is conventionally assumed a neutral tracer in the variable K_d model.

Reactive transport problems are usually highly demanding in terms of computational resources. More than 90% of computational time is normally spent on chemical equilibrium calculations. Thus, parallelization of these models is of major importance. Computation of the reactive transport problem described above was performed on cluster [6] of nodes with Intel Xeon X5650@2.67 GHz processors. Performance results are shown in Table 4.7, where S_2 shows the speedup with respect to the initial

Table 4.7 Parallel efficiency of the reactive transport model

No. of cores	Computational time (s)	Efficiency S_2	Efficiency $S_{p/2}$
2	86215	1.00	–
4	41767	2.06	2.06
8	25631	3.36	1.62
16	15313	5.63	1.67
32	9283	9.28	1.64
64	4697	18.35	1.97
128	2982	28.90	1.57

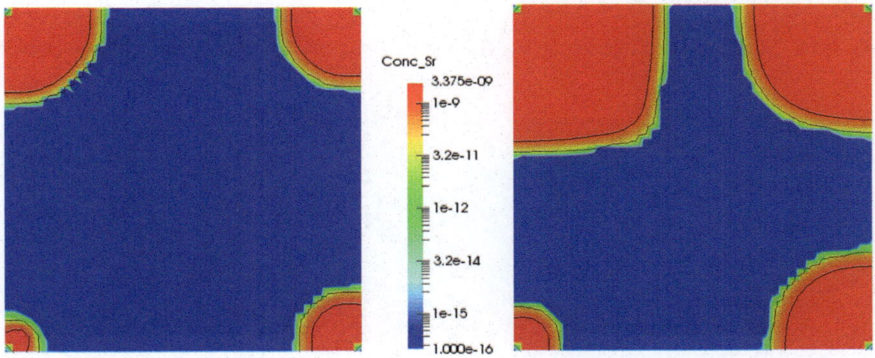

Fig. 4.14 Strontium propagation at T = 1000 years (left) and T = 3000 years (right), obtained using reactive transport model

computation on 2 cores and $S_{p/2}$ is the speedup achieved with respect to computation on twice less number of cores.

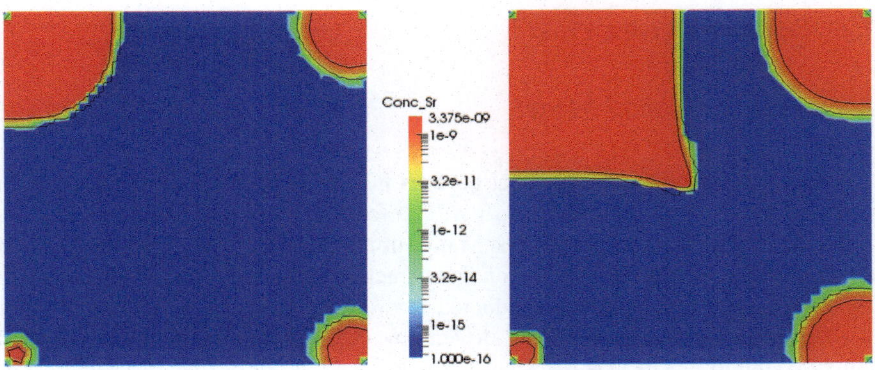

Fig. 4.15 Strontium propagation at T = 1000 years (left) and T = 3000 years (right), obtained using variable K_d model

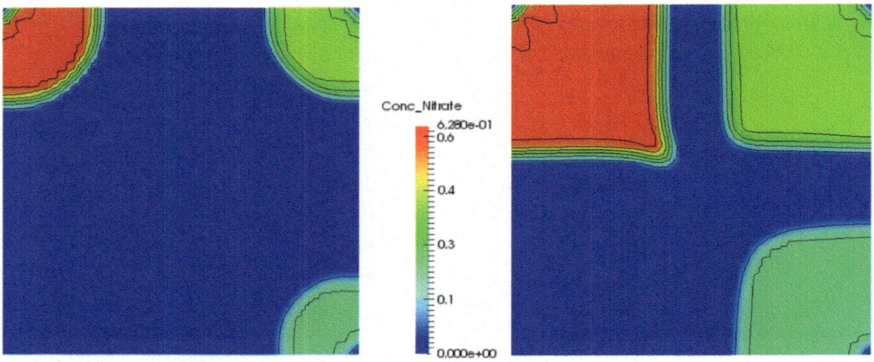

Fig. 4.16 Nitrate ion propagation at T = 1000 years (left) and T = 3000 years (right), obtained using reactive transport model

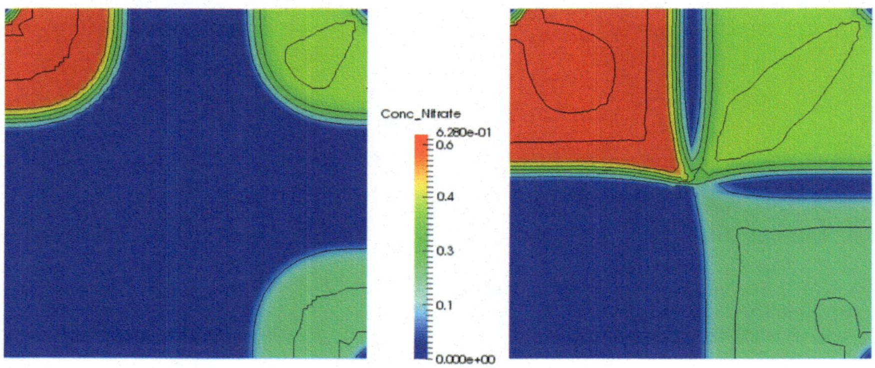

Fig. 4.17 Nitrate ion propagation at T = 1000 years (left) and T = 3000 years (right), obtained using variable K_d model

4.4 Density-Driven Flow

4.4.1 Mathematical Model

The considered mathematical model governs flow of a variable density solution in saturated or unsaturated conditions. It will be formulated in terms of mass concentration and for isothermal conditions. Many simulation codes use the formulation of solute transport equation in terms of mass fraction instead of mass concentration. Experimental comparison was conducted in order to assess the influence of these two different approaches on density-driven flow (DDF) modeling. Kolditz et al. [95] compared the ROCKFLOW (uses mass fraction) and the FEFLOW (uses mass concentration) codes on the standard Henry [81], Elder [61, 150], and salt dome [24] benchmarks. Good agreement between the results of two tested codes was obtained. Younes et al. [154] applied their code TVDV-2D to the salt dome test case using both formulations and concluded that the obtained results were very close and that the simpler formulation in terms of mass concentration might be used. Thus, we can conclude that the use of the simpler approximation is justified and widely accepted. Assuming only one contaminant present, the system of flow and transport equations is the following:

$$\frac{\partial (S\varphi\rho)}{\partial t} + \text{div}\,(\rho\mathbf{q}) = Q\rho_s, \tag{4.43}$$

$$\frac{\partial ((S\varphi + \rho_b K_d)\, c)}{\partial t} + \text{div}\,(\mathbf{q}c) - \text{div}\,\mathbb{D}\nabla c = Qc_s. \tag{4.44}$$

Here, ρ is the solute density, Q is the source term, and c_s and ρ_s are contaminant concentration and density of the solution in the source. The choice of mass concentration as the primary variable is done first of all to avoid the need to solve a nonlinear transport equation. It allows us to express the contaminant mass conservation law linearly in terms of mass concentration, otherwise we obtain a nonlinear term stemming from the diffusion operator, where density depends on mass fraction. Younes et al. [154] proposed an efficient way to reduce the nonlinear diffusion equation to a linear one in terms of squared density using the advective form of the transport equation, but we do not know such an elegant solution for the divergence form, which is preferred for the finite volume discretization. Another reason to use Eq. (4.44) is the ability to apply the same transport solver which had been used for transport problems with constant density.

The system of equations (4.43), (4.44) must be supplemented by constitutive relations. First, it is the generalized Darcy's law for the dependence of Darcy velocity on pressure head and density:

$$\mathbf{q} = -\frac{\mu_0}{\mu}\mathbb{K}\left(\nabla\psi_0 + \frac{\rho}{\rho_0}\nabla z\right), \tag{4.45}$$

where ρ_0 is the reference density of the liquid in the absence of contaminants (freshwater density), μ and μ_0 are dynamic viscosities of the solute and of freshwater, respectively, z is the vertical coordinate, $\psi_0 = \frac{p_w}{\rho_0 g}$ is the pressure head measured in terms of reference liquid column height, and g is the gravity constant. Dynamic viscosity $\mu(c)$ is assumed to be a function of concentration.

Second, density of the liquid depends on contaminant concentration. We assume a linear dependence:

$$\rho = \rho_0 + \varkappa_{\text{vol}} c, \tag{4.46}$$

where, $\varkappa_{\text{vol}}$ is the volumetric expansion coefficient. Influence of pressure on density is neglected.

When modeling density-driven flow in confined conditions, the flow equation is normally written in terms of equivalent freshwater head $h_0 = \frac{p}{\rho_0 g} + z$

$$\rho s_0 \frac{\partial \mathrm{h}_0}{\partial t} + \varphi \varkappa_{\mathrm{vol}} \frac{\partial c}{\partial t} - \operatorname{div}\left(\rho \frac{\mu_0}{\mu} \mathbb{K}\left(\nabla \mathrm{h}_0 + \frac{\rho - \rho_0}{\rho_0} \nabla z\right)\right) = Q \rho_s \tag{4.47}$$

and the transport equation can be written in the form (4.2).

When dealing with the radioactive waste, volumetric heat generation due to radioactive decay and the corresponding thermal convection might be significant. The corresponding mathematical model is implemented in GeRa and its verification can be found in [73].

4.4.2 Numerical Solution Aspects

Similar to reactive transport problems (see Sect. 4.3), DDF problems may be solved using different approaches for coupling: fully coupled with Newton iterations, fixed-point iteration, and operator-splitting. Here, we focus on the fixed-point iterative scheme, allowing for sequential separate solution of flow and transport problems. Its advantage is the ease of use of existing flow and transport solvers and to apply the solvers better fitted for each of these problems. Let k be the iteration count and n be the time step number. First, we formulate the backward Euler implicit scheme for the system (4.43), (4.44):

$$\rho^{n+1} \frac{\theta^{n+1} - \theta^n}{\Delta t} + S^{n+1} \rho^{n+1} s_0 \frac{\psi^{n+1} - \psi^n}{\Delta t} + \theta^n \varkappa_{\mathrm{vol}} \frac{C^{n+1} - C^n}{\Delta t} + \\ + \operatorname{div}\left(\mathbf{q}^{n+1} \rho^{n+1}\right) = Q^{n+1} \rho_s^{n+1}, \tag{4.48}$$

$$\frac{\theta^{n+1} C^{n+1} - \theta^n C^n}{\partial t} + \operatorname{div}\left(\mathbf{q}^{n+1} C^{n+1}\right) - \operatorname{div} \mathbb{D} \nabla C^{n+1} = Q^{n+1} C_s^{n+1}. \tag{4.49}$$

We should remark here that in case of sources (with $Q > 0$) the total mass of the injected solution and contaminant during the time step should be put into the right-hand sides of (4.48), (4.49) in order to conserve mass. In case of sinks ($Q < 0$), the values ρ^{n+1} and C^{n+1} are used, respectively. Here, it is assumed that sources' and sinks' intensity Q does not depend on unknowns ψ and c. The fixed-point iteration algorithm for this system is implemented in the following way:

1. $$\rho^{n+1,k} \frac{\theta^{n+1,k+1} - \theta^n}{\Delta t} + S^{n+1,k+1} \rho^{n+1,k} s_0 \frac{\psi^{n+1,k+1} - \psi^n}{\Delta t} + \\ + \theta^n \varkappa_{\mathrm{vol}} \frac{C^{n+1,k} - C^n}{\Delta t} + \nabla\left(\rho^{n+1,k} \mathbf{q}^{n+1,k+1}\right) = Q^{n+1} \rho_s^{n+1,k}, \tag{4.50}$$

2. $$\frac{\theta^{n+1,k+1} C^{n+1,k+1} - \theta^n C^n}{\partial t} + \nabla\left(C^{n+1,k+1} \mathbf{q}^{n+1,k+1}\right) - \\ - \nabla \cdot \mathbb{D} \nabla C^{n+1,k+1} = Q^{n+1} C_s^{n+1,k+1}. \tag{4.51}$$

The flow problem (4.50) is solved using Picard or Newton iterative methods (see Sect. 4.2). When integrating over a cell ω to apply the FV method for space discretization, the advection term is represented as a surface integral:

$$\int_{\omega} \nabla\left(\rho^{n+1,k}\mathbf{q}^{n+1,k+1}\right) d\omega = -\int_{\partial\omega} \rho_{upw}^{n+1,k}\frac{\mu_0}{\mu}\mathbb{K}\left(\nabla\psi_0^{n+1,k+1} + \frac{\rho_{av}^{n+1,k}}{\rho_0}\nabla z\right)\cdot \mathbf{n} dS. \tag{4.52}$$

As concentration is approximated in the FV schemes via a piecewise-constant function, density is also assigned a constant value in each grid cell, and therefore is discontinuous on faces. In order to evaluate the surface integral in (4.52), one has to define density on cell faces. Similar to [74], we use averaged density value ρ_{av} (half of the sum of densities from two neighboring cells) within the flux term and upwind value ρ_{upw} for the density out of the flux term. The convergence criterion for the iterative process (4.48), (4.49) is that the relative (with respect to the norm of the variables vector) change in primary variables is below a specified value.

4.4.3 Numerical Examples

4.4.3.1 The Elder Problem

The Elder problem is one of the most widely used "benchmarks" for DDF numerical models. The history of this test lasts for more than 50 years [63] and starts from the experiment of Elder [62] on thermal convection in a Hele-Shaw cell and the corresponding numerical simulation [61, 62]. Later, the test was reformulated as an equivalent DDF problem in [150]. Numerous codes were tested on the Elder problem, namely, FEFLOW and ROCKFLOW [95], SUTRA [152], TOUGH2 [118], and many others (see, e.g., [18, 69]). The simulation results appeared to be contradictory. Using different codes and different grids, two major solution patterns were obtained: first, with two convective cells and downwelling flow in the center and, second, with four convective cells and upwelling central flow. In some works, even three solution patterns were found [69]. Finally, it was confirmed that multiple steady states exist for the Elder problem [147].

The computational domain is a rectangle 600×150 m (see Fig. 4.18). Dirichlet boundary conditions for transport are set on a part of the top boundary (dimensionless concentration $c = 1$, meaning maximum salt concentration) and on the whole bottom boundary ($c = 0$). For the flow equation, impermeable boundary conditions ($\mathbf{q}\cdot\mathbf{n} = 0$ m/day) are set on all the boundaries except for the two top corners with Dirichlet boundary conditions $\mathrm{h}_0 = 0$ m. Media and solute parameters are stated in Table 4.8.

The results of simulation in GeRa on a 60×25 grid using two different schemes for the transport equation discretization are shown in Figs. 4.19, 4.20. Isochlores of relative concentrations 0.2, 0.4, 0.6, and 0.8 are drawn. These flow patterns correspond to the most commonly met ones in literature.

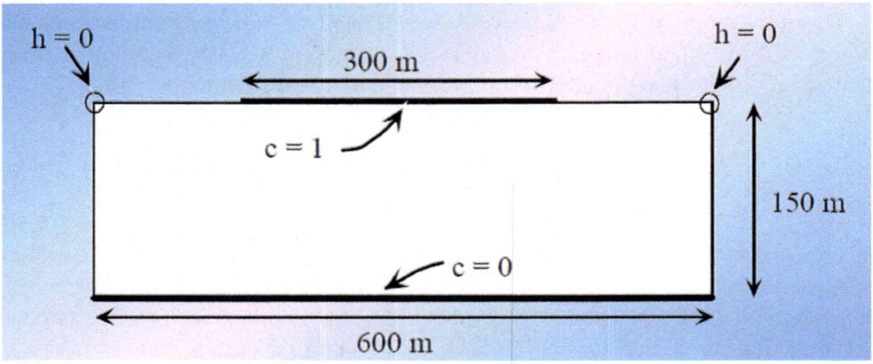

Fig. 4.18 The Elder problem geometry and boundary conditions

Table 4.8 Native solution parameters

Parameter	Value
Molecular diffusion coefficient, m^3/day	0.308
Hydraulic conductivity, m/day	0.411
Fresh/saline(maximum) water density, /m^3	1000/1200
Porosity	0.1
Salt concentration corresponding to saline water (maximum), kg/m^3	285.7

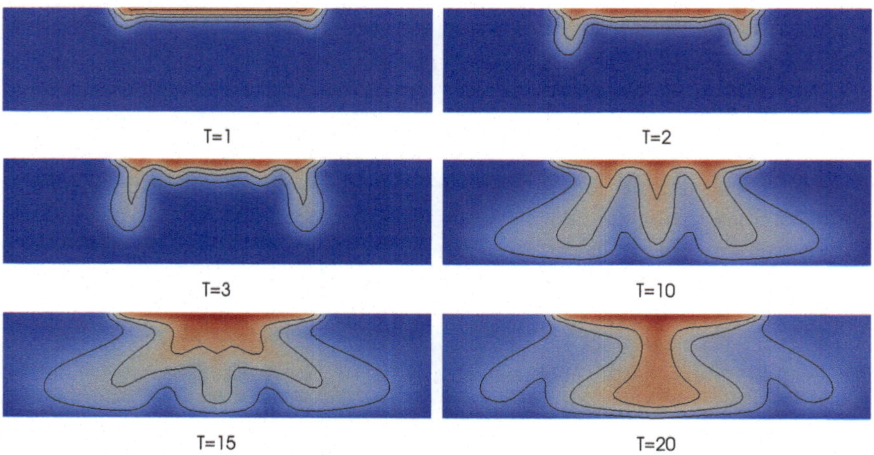

Fig. 4.19 The Elder problem calculation results using full implicit first-order upwind FV scheme

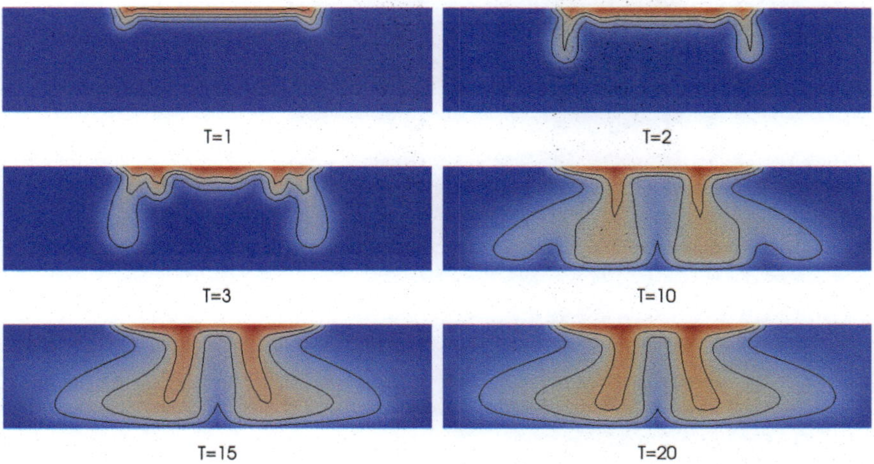

Fig. 4.20 The Elder problem calculation results using explicit for convection second-order TVD and implicit for diffusion FV scheme

There is one remarkable feature of the Elder problem which is often met in practical applications. On the dissolution boundary, a combination of Dirichlet boundary condition for transport is combined with zero-flow condition for flow. Thus, the salt mass flux is taken into account only in the transport equation but is neglected in the flow equation. The problem of proper dissolution boundary conditions formulation was first studied by Hassanizadeh and Leijnse [80]. They proposed a consistent form of the flow boundary condition on dissolution boundary Γ, here formulated in terms of mass concentration:

$$\mathbf{q} \cdot \mathbf{n}|_{\Gamma} = \frac{\mathbf{J} \cdot \mathbf{n}}{1 - c_{sat}/\rho_{sat}}, \tag{4.53}$$

where c_{sat} and ρ_{sat} are the concentration and density at maximum salinity, and $\mathbf{J}$ is the diffusion flux. Hidalgo et al. [83] have shown that inconsistent boundary conditions imposing causes unfeasible pressure drop and qualitatively wrong flow problem calculation. For the Elder problem, this leads to inflow of pure water from the corner points instead of water outflow due to the growth of fluid density and volume because of salt dissolution. Calculations in GeRa show that inconsistent boundary conditions lead to a total inflow of 538 m^3 of water into the domain which is 5 m thick (the extra dimension added to use the 3D code for the 2D problem), see Fig. 4.21.

4.4.3.2 Injection of Dense Brine into a Tilted Aquifer

This test is designed to assess the influence of DDF on the propagation of liquid radioactive waste (LRW) injected into deep tilted aquifer. The design of existing LRW injection sites [111, 127] presumes codirectional regional flow field and elevation of

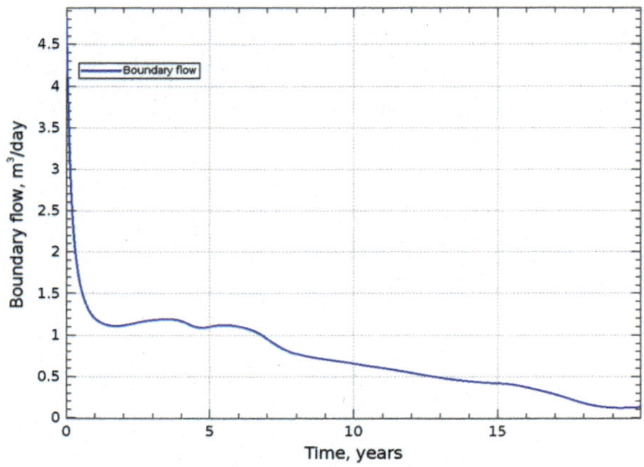

Fig. 4.21 Intensity of water inflow through Dirichlet-type boundaries for the Elder problem

Table 4.9 Parameters of the problem

Parameter	Value
Aquifer thickness, m	50
Tangent of the aquifer sloping angle	0.028
Volumetric expansion coefficient	0.6
Contaminant concentration in injected solution, g/l	250
Hydraulic conductivity, m/day	1
Porosity	0.2
Longitudinal dispersivity, m	0.1
Transversal dispersivity, m	0.01

the aquifer bottom. Thus, the buoyancy forces applied to injected dense brines can suppress the regional flow and significantly retard the pollution propagation or even lead to trapping of the brine in a natural cavern [110]. In [110, 111], the problem was studied theoretically and numerically in 2D using Boussinesq assumption and in the assumption of a thin aquifer. As the Boussinesq assumption is applicable only for the case of low contaminant concentrations, and in case of LRW polygons the salinity reaches 240–250 g/l and can not be treated as low, we find it useful to perform a 3D simulation taking into account the real thickness of the aquifer. Problem parameters are taken mainly from [111] and given in Table 4.9. The computational domain is a square [−2000 m; 2000 m] × [−2000 m; 2000 m] in the horizontal plane; the injection well is located in the middle of it and its filter covers all the aquifer. We assume that

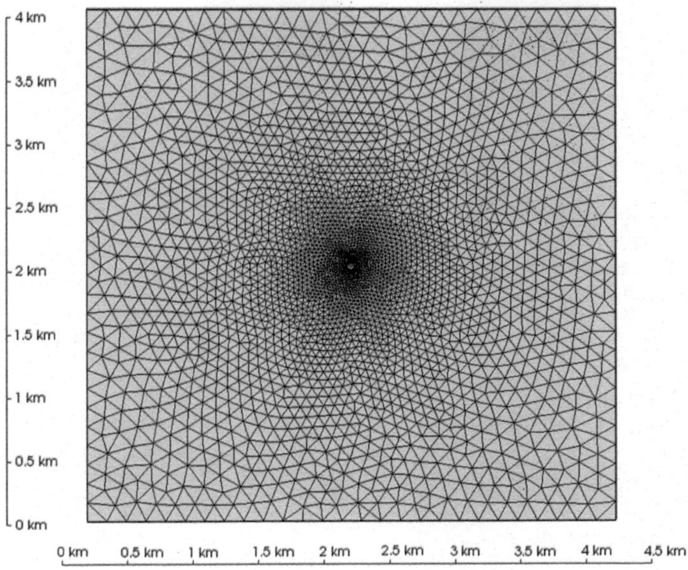

Fig. 4.22 Footprint of the computational grid

the influence of the well on the domain boundaries is negligible. The regional flux and the aquifer elevation are both directed along the x-axis. Boundary conditions on the top and bottom and on the sides of the domain parallel to the y-axis are set as impermeable for flow. Piezometric head is prescribed on the boundaries $= -2000$ m ($h = 1.28$ m) and $x = 2000$ m ($h = 0$ m) in order to obtain Darcy velocity 3.2E-4 m/day.

The schedule of the injection is the following: during 50 years every year, the injection cycle is performed during 60 days with a debit of 300 m^3/day and contaminant concentration 250 g/l, after that injection is stopped. Modeling is performed using the fully implicit FV scheme on a grid adapted to the well composed of triangular prisms and containing 110 thousand cells with 20 vertical cell layers (see Fig. 4.22). The time period is 200 years; the time step is 30 days.

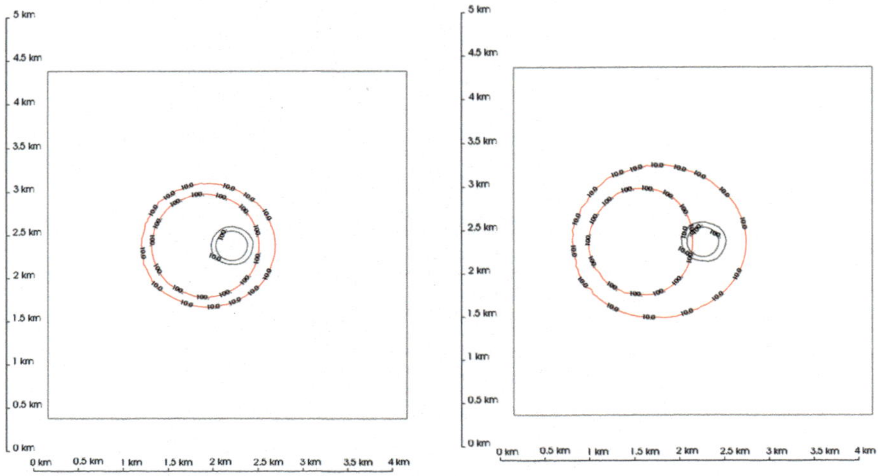

Fig. 4.23 Isolines of contaminant concentration on the aquifer bottom at T = 100 years (left) and T = 200 years (right): black—without DDF, red—with DDF

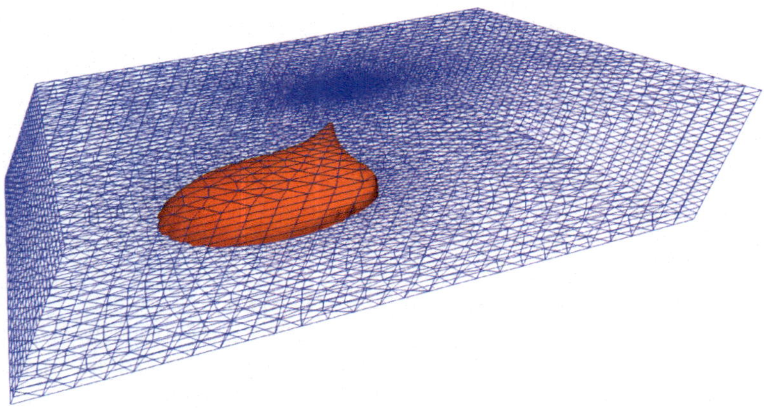

Fig. 4.24 Contaminant plume $c > 1$ g/l at T = 100 years (scaled 20 times vertically)

Simulation results are shown in Figs. 4.23 and 4.24. One can see that dense solute due to gravity sinks onto the bottom of the aquifer. This effect leads to much larger spreading of the contamination plume within the aquifer, which could not be observed in 2D models. Even though at T = 200 years (see Fig. 4.23) one can see that the most dense solution has moved downdip the aquifer, the less dense solution is likely to reach the discharge area even faster than in case of no DDF effects.

Chapter 5
Application of MFV in Modeling of Coagulation of Blood Flow

The purpose of blood flow modeling and multi-physics modeling of coagulation of blood flow is to predict thromboembolism in blood vessels and heart chambers. Cardiovascular diseases are the main causes of human death in the world. Thromboembolitic complications are one of the main reasons for strokes and infarcts. A personalized model allows one to predict clot growth under different kinds of medications and to assess the risk of blood vessel occlusion leading to new healthcare practices. In this chapter, we address our FV model for coagulation of blood flow validated by laboratory experiments.

5.1 Model of Blood Flow and Coagulation

The dynamics of the blood plasma occupying a domain Ω is described by the incompressible Navier–Stokes equations with a hydraulic permeability term:

$$\begin{aligned} \frac{\partial \rho \mathbf{u}}{\partial t} + \operatorname{div}\left(\rho \mathbf{u}\mathbf{u}^T - \mu \nabla \mathbf{u} + p\mathbb{I}\right) &= -\frac{\mu}{K_f}\mathbf{u}, \\ \operatorname{div}(\mathbf{u}) &= 0. \end{aligned} \tag{5.1}$$

Here, the velocity and the pressure of the blood plasma are denoted by $\mathbf{u}$ and p, respectively. The density ρ and viscosity μ of blood plasma are assumed to be constant. In Eq. (5.1), the parameter K_f is the hydraulic permeability of a porous clot composed from fibrin polymer. The clot may grow according to blood biochemistry. The right-hand side of (5.1) couples the blood flow with the blood coagulation process. A model of the coagulation process describes clot formation using thefollowing

Y. Vassilevski et al., *Parallel Finite Volume Computation on General Meshes*,
https://doi.org/10.1007/978-3-030-47232-0_5

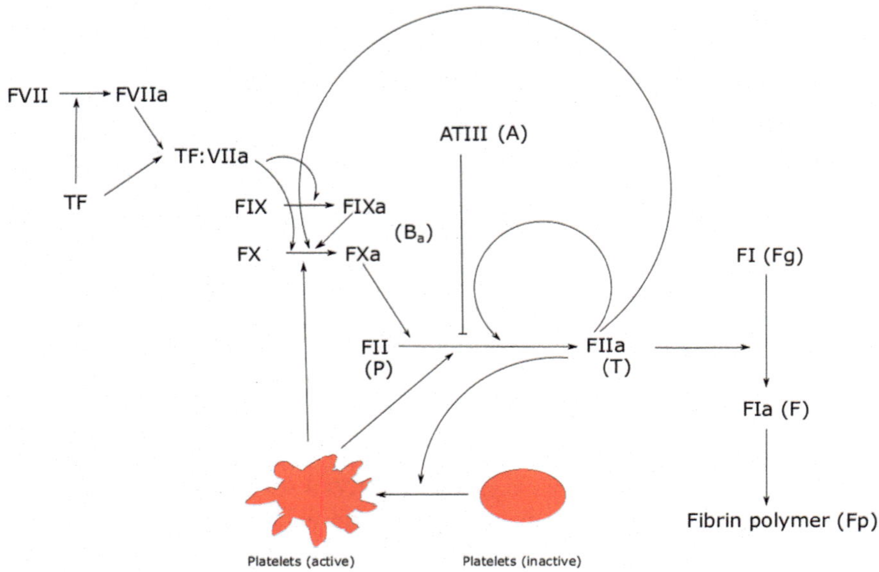

Fig. 5.1 Graphic illustration of reaction cascade for blood clotting process (5.2)

concentrations of reactive blood components: prothrombin PT, thrombin T, antithrombin A, blood clotting factor FX B_α, fibrin F, fibrinogen F_g, fibrin polymer F_p, free platelets ϕ_f in the flow, and platelets ϕ_c trapped in the clot.

The hydraulic permeability depends on the normalized concentration $\hat{F}_p = \min\left(\frac{7}{10}, \frac{F_p}{7000}\right)$ of fibrin polymer in the clot and concentration of platelets ϕ_c trapped in the clot:

$$\frac{1}{K_f} = \frac{16}{a^2}\hat{F}_p^{\frac{3}{2}}\left(1 + 56\hat{F}_p\right)\frac{\phi_{max} + \phi_c}{\phi_{max} - \phi_c},$$

where a is the radius of the fibers in the fibrin polymer [153].

A simplified reaction cascade of the coagulation model is shown in Fig. 5.1.

The model [38] consists of unsteady reaction-advection–diffusion equations for concentrations of all reactive components except fibrin polymer and platelets. Fibrin polymer satisfies a simple reaction equation, whereas the platelets follow the nonlinear advection–diffusion of traffic flow. The model of the coagulation is

$$
\begin{aligned}
&\frac{\partial P}{\partial t}+\operatorname{div}(P\mathbf{u}-D\nabla P)=-\left(k_1\phi_c+k_2B_\alpha+k_3T+k_4T^2+k_5T^3\right)P,\\
&\frac{\partial T}{\partial t}+\operatorname{div}(T\mathbf{u}-D\nabla T)=\left(k_1\phi_c+k_2B_\alpha+k_3T+k_4T^2+k_5T^3\right)P-k_6AT,\\
&\frac{\partial B_\alpha}{\partial t}+\operatorname{div}(B_\alpha\mathbf{u}-D\nabla B_\alpha)=(k_7\phi_c+k_8T)\left(B^0-B_\alpha\right)-k_9AB_\alpha,\\
&\frac{\partial A}{\partial t}+\operatorname{div}(A\mathbf{u}-D\nabla A)=-k_6AT-k_9AB_\alpha,\\
&\frac{\partial F_g}{\partial t}+\operatorname{div}\left(F_g\mathbf{u}-D\nabla F_g\right)=-\frac{k_{10}TF_g}{K_{10}+F_g},\\
&\frac{\partial F}{\partial t}+\operatorname{div}(F\mathbf{u}-D\nabla F)=\frac{k_{10}TF_g}{K_{10}+F_g}-k_{11}F,\\
&\quad\frac{\partial F_p}{\partial t}=k_{11}F,\\
&\frac{\partial\phi_c}{\partial t}+\operatorname{div}\left(\tanh\left(\pi\left(1-\frac{\phi_c+\phi_f}{\phi_{max}}\right)\right)(\phi_c\mathbf{u}-D_p\nabla\phi_c)\right)=-(k_{12}T-k_{13}\phi_c)\phi_f,\\
&\frac{\partial\phi_f}{\partial t}+\operatorname{div}\left(\tanh\left(\pi\left(1-\frac{\phi_c+\phi_f}{\phi_{max}}\right)\right)(\phi_f\mathbf{u}-D_p\nabla\phi_f)\right)=(k_{12}T-k_{13}\phi_c)\phi_f.
\end{aligned}
\tag{5.2}
$$

The multi-physics model (5.1)–(5.2) combines the model of blood plasma flow and the model of the simplified coagulation cascade, assuming that

- blood plasma is incompressible Newtonian fluid, and complex rheology of blood is ignored;
- polymerized fibrin does not move with the flow, i.e., the clot cannot detach and travel with the flow;
- the blood vessel boundaries are non-porous and rigid, and their poro-elasticity is ignored.

The set of kinetic reactions between components of the flow and parameters of these reactions are taken from [38, 39]. Some parameters were obtained from literature; the other parameters were identified using a thrombin-generation model or were fitted to the experiment of blood plasma flow through microvessels [134].

We split the boundary $\partial\Omega$ into slip boundary Γ_{slip} and inflow/outflow boundary Γ_{pres} and use slip boundary condition on Γ_{slip} and pressure drop boundary condition on Γ_{pres}, see (2.92) and Table 2.1. For Eqs. (5.2), there may be three types of boundary conditions, depending on the blood component. Most of components satisfy Dirichlet-type conditions $C = C_b$ on Γ_D or Neumann boundary conditions $\nabla C = \mathbf{0}$ on Γ_N. The blood clotting factor FX B_α on the injured part of blood vessel Γ_I (which is the source of tissue factor) satisfies a specific boundary condition

$$
\frac{\partial B_\alpha}{\partial t}=\frac{\alpha(B^0-B_\alpha)}{1+\beta(B^0-B_\alpha)}\ \text{on}\ \Gamma_I,
\tag{5.3}
$$

with given parameters α and β.

5.2 FV Discretization of Blood Coagulation Model

All the unknowns (velocity vector $\mathbf{u}$, pressure p and all the blood components $\Theta = (P, T, B_\alpha, A, F_g, F, F_p, \phi_c, \phi_f)^T$) are collocated at the centers of the cells. The cell-centered finite volume discretization results in a system composed of 13 unknowns and Eqs. (5.1)–(5.2) per computational cell. The whole system is solved simultaneously in a semi-implicit manner by the backward Euler time-stepping scheme. The only source of explicitness is an extrapolation in time for reactions, as discussed below.

FV discretization of the Navier–Stokes equations (5.1) is discussed in Sect. 2.7. Below, we present FV discretization of the reaction-advection–diffusion equations (5.2).

Consider an equation from (5.2) for a component C other than platelets. It contains diffusion flux, advection flux, and reaction terms. The diffusion flux $-D\nabla C \cdot \mathbf{n}$ is discretized by the nonlinear TPFA from Sect. 2.3. The advection flux $\mathbf{n}^T\mathbf{u}C$ is discretized by the first-order upwind method (2.63). The positivity-preserving property of both schemes is very important for stability of the reactions.

The Reaction terms are formed by the right-hand side of (5.2) and are denoted by vector $R(\Theta)$. The reactions do not need discretization in space. Contributions of some blood components into reaction expression are extrapolated from previous time steps. The purpose of the extrapolation is to eliminate large contributions to off-diagonal terms in the Jacobian. The extrapolated terms were picked by trial-and-error during Jacobian's analysis of the 0D model of the reactions with various choices of the initial state Θ^0.

The following extrapolation-based approximation to $R(\Theta^{n+1})$ was obtained:

$$R(\Theta^{n+1}) \approx \begin{pmatrix} -\left(k_1\hat{\phi}_c + k_2 B_\alpha^{n+1} + k_3\hat{T} + k_4\hat{T}^2 + k_5\hat{T}^3\right) P^{n+1} \\ \left(k_1\hat{\phi}_c + k_2 B_\alpha^{n+1} + k_3\hat{T} + k_4\hat{T}^2 + k_5\hat{T}^3\right) P - k_6 A^{n+1} T^{n+1} \\ \left(k_7\phi_c^{n+1} + k_8 T^{n+1}\right)\left(B^0 - B_\alpha^{n+1}\right) - k_9 A^{n+1} B_\alpha^{n+1} \\ -\left(k_6 T^{n+1} + k_9 B_\alpha^{n+1}\right) A^{n+1} \\ -\frac{k_{10}\hat{T} F_g^{n+1}}{K_{10}+F_g^{n+1}} \\ \frac{k_{10}\hat{T} F_g^{n+1}}{K_{10}+F_g^{n+1}} - k_{11} F^{n+1} \\ k_{11} F^{n+1}, \\ -\left(k_{12}\hat{T} - k_{13}\phi_c^{n+1}\right)\phi_f^{n+1} \\ \left(k_{12}\hat{T} - k_{13}\phi_c^{n+1}\right)\phi_f^{n+1} \end{pmatrix}, \tag{5.4}$$

where $\hat{T} = 2T^n - T^{n-1}$ and $\hat{\phi}_c = 2\phi_c^n - \phi_c^{n-1}$ are extrapolations in time for T^{n+1} and ϕ_c^{n+1}, respectively. These are the only explicit terms in otherwise fully implicit time integration for the multi-physics model (5.1)–(5.2). Such semi-implicit scheme remains stable even for very large time steps Δt, whereas the fully implicit model (with $\hat{T} = T^{n+1}$ and $\hat{\phi}_c^{n+1} = \phi_c^{n+1}$) struggles with large time steps because of large off-diagonal terms in the Jacobian. A general stable approach to approximation of $R(\Theta)$ based on analysis of eigenvalues of its Jacobian is possible; however, it is more expensive computationally, see Remark 5.1.

It remains to discuss discretization of the nonlinear advection–diffusion of free platelets and platelets trapped in the clot. The two equations are coupled through the reactions and the nonlinear coefficient $\tanh\left(\pi\left(1-\frac{\phi_c+\phi_f}{\phi_{max}}\right)\right)$.

First, we consider the nonlinearity in the advection flux of the form $\lambda(C)\mathbf{n}^T\mathbf{u}$. Different functions $\lambda(C)$ produce different types of advection problems:

- simple advection for $\lambda(C) = C$,
- traffic advection for $\lambda(C) = C(1-C)$,
- platelets advection $\lambda(C) = C\tanh(\pi(1-C))$.

Traffic advection and platelets advection have non-monotone function $\lambda(C)$ whose derivative changes sign. Therefore, the first-order upwind scheme should account the sign of $\lambda'(C)$.

To be more precise, we consider two concentrations C_+, C_- collocated in cells ω_+, ω_- which share a face f. Discretization of a nonlinear advective flux is stable for all differentiable functions $\lambda(C)$ when the following conditions are satisfied:

- diagonal entries of the Jacobian matrix for the flux are positive and off-diagonal entries are non-positive;
- if both ω_+ and ω_- yield admissible fluxes for contribution to the Jacobian matrix, then the smallest flux value should be chosen[1];
- if both ω_+ and ω_- yield non-admissible fluxes for contribution to the Jacobian matrix, then the flux should be chosen so that it will not contribute to the Jacobian matrix.[2]

To cope with the above requirements, we define the following upwind discretization of a nonlinear advection flux:

$$\lambda(C)\mathbf{n}^T\mathbf{u} = \begin{cases} \lambda(C_+)\mathbf{n}^T\mathbf{u}, & \lambda'(C_+)\mathbf{n}^T\mathbf{u} \geq 0,\ \lambda'(C_-)\mathbf{n}^T\mathbf{u} \geq 0, \\ \lambda(C_-)\mathbf{n}^T\mathbf{u}, & \lambda'(C_+)\mathbf{n}^T\mathbf{u} < 0,\ \lambda'(C_-)\mathbf{n}^T\mathbf{u} < 0, \\ \mathrm{minmod}\left(\lambda(C_+), \lambda(C_-)\right)\mathbf{n}^T\mathbf{u}, & \lambda'(C_+)\mathbf{n}^T\mathbf{u} \geq 0,\ \lambda'(C_-)\mathbf{n}^T\mathbf{u} < 0, \\ \lambda(C^*)\mathbf{n}^T\mathbf{u},\ \lambda'(C^*) = 0, & \lambda'(C_+)\mathbf{n}^T\mathbf{u} < 0,\ \lambda'(C_-)\mathbf{n}^T\mathbf{u} \geq 0. \end{cases} \tag{5.5}$$

Second, we address coupling of the two nonlinear advection–diffusion equations for free and trapped platelets. Following the holistic approach introduced in Chap. 2, we consider the flux vector for both equations on face f between two neighboring cells ω_+ and ω_-:

---

[1]This avoids concentration overflow in the acceptor cell or underflow in the donor cell.

[2]In case of the traffic advection, this corresponds to traffic motionlessness for the moment of sudden change of the red light to the green one.

$$\mathbf{t}(\phi_c, \phi_f) = \begin{pmatrix} t_1(\phi_c, \phi_f) \\ t_2(\phi_c, \phi_f) \end{pmatrix} = \tanh\left(\pi\left(1 - \frac{\phi_c + \phi_f}{\phi_{max}}\right)\right) \mathbf{n}^T \left(\mathbf{u} - D_p \nabla\right) \begin{pmatrix} \phi_c \\ \phi_f \end{pmatrix}. \tag{5.6}$$

Variation of the flux is written with the help of its Jacobian $Q(\phi_c, \phi_f)$:

$$\begin{pmatrix} \mathrm{d}t_1(\phi_c, \phi_f) \\ \mathrm{d}t_2(\phi_c, \phi_f) \end{pmatrix} = \begin{pmatrix} \frac{\partial t_1(\phi_c,\phi_f)}{\partial \phi_c} & \frac{\partial t_1(\phi_c,\phi_f)}{\partial \phi_f} \\ \frac{\partial t_2(\phi_c,\phi_f)}{\partial \phi_c} & \frac{\partial t_2(\phi_c,\phi_f)}{\partial \phi_f} \end{pmatrix} \begin{pmatrix} \mathrm{d}\phi_c \\ \mathrm{d}\phi_f \end{pmatrix} := Q(\phi_c, \phi_f) \begin{pmatrix} \mathrm{d}\phi_c \\ \mathrm{d}\phi_f \end{pmatrix}. \tag{5.7}$$

Let M_+ and M_- be two matrices of order 2 (to be defined later) such that $M_+ + M_- = I$. Then for the combined argument of platelets concentrations in cells ω_+, ω_-

$$\Phi = M_+ \begin{pmatrix} \phi_{c,+} \\ \phi_{f,+} \end{pmatrix} + M_- \begin{pmatrix} \phi_{c,-} \\ \phi_{f,-} \end{pmatrix}, \tag{5.8}$$

we write the flux vector variation

$$\mathrm{d}\mathbf{t}(\Phi) = Q(\Phi) M_+ \begin{pmatrix} \mathrm{d}\phi_{c,+} \\ \mathrm{d}\phi_{f,+} \end{pmatrix} + Q(\Phi) M_- \begin{pmatrix} \mathrm{d}\phi_{c,-} \\ \mathrm{d}\phi_{f,-} \end{pmatrix}. \tag{5.9}$$

We recall that we use the normal font for matrix coefficients (M, Q and its decompositions) associated with flux.

For the sake of stable approximation, matrices M_+ and M_- should be chosen so that $Q(\Phi)M_+$ has positive eigenvalues and $Q(\Phi)M_-$ has negative eigenvalues. To this end, we use Algorithm 5.1.

Algorithm 5.1 Eigendecomposition-based iterations for matrices M_+ and M_-

1: set $M_+^0 = M_-^0 = \frac{1}{2} I$
2: **for** $k = 0, 1, \dots$ **do**
3: $\quad \Phi^k = M_+^k \begin{pmatrix} \phi_{c,+} \\ \phi_{f,+} \end{pmatrix} + M_-^k \begin{pmatrix} \phi_{c,-} \\ \phi_{f,-} \end{pmatrix}$
4: $\quad Q(\Phi^k) = L \Lambda L^{-1}$ ▷ eigendecomposition
5: $\quad M_+^k = \frac{1}{2} L \left(\mathrm{sgn}(\Lambda) + |\mathrm{sgn}(\Lambda)|\right) L^{-1}$
6: $\quad M_-^k = \frac{1}{2} L \left(\mathrm{sgn}(\Lambda) - |\mathrm{sgn}(\Lambda)|\right) L^{-1}$
7: $\quad$ **if** $\|Q(\Phi^k) - Q(\Phi^{k-1})\| < \varepsilon$ **then**
8: $\qquad$ stop iterate
9: $\quad$ **end if**
10: **end for**

Typically, in complicated cases, the algorithm converges in 3–5 iterations for $\varepsilon = 10^{-4}$. The eigendecomposition in Algorithm 5.1 may not exist or may have complex eigenvalues. In both cases, the iterations are not applicable, and instead of eigendecomposition one applies the SVD decomposition $Q(\Phi^k) = U \Sigma V^T$. Let u_i and v_i represent the ith column of U and V, respectively, and σ_i be the ith singular value. If $u_i \cdot v_i < 0$, we flip the signs for σ_i and v_i: $\hat{\sigma}_i = -\sigma_i$, $\hat{v}_i = -v_i$, and define the alternative SVD decomposition $Q(\Phi^k) = U\hat{\Sigma}\hat{V}^T$. The alternative SVD decomposition coincides with eigendecomposition for symmetric matrices; otherwise, it differs. Iterations of Algorithm 5.1 transform to SVD-based iterations:

Algorithm 5.2 SVD-based iterations for matrices M_+ and M_-

```
set M_+^0 = M_-^0 = (1/2) I
for k = 0, 1, ... do
    Φ^k = M_+^k (φ_{c,+}; φ_{f,+}) + M_-^k (φ_{c,-}; φ_{f,-})
    Q(Φ^k) = U Σ̂ V̂^T                                   ▷ modified SVD
    M_+^k = (1/2) V̂ (sgn(Σ̂) + |sgn(Σ̂)|) V̂^T
    M_-^k = (1/2) V̂ (sgn(Σ̂) − |sgn(Σ̂)|) V̂^T
    if ||Q(Φ^k) − Q(Φ^{k−1})|| < ε then
        stop iterate
    end if
end for
```

Once the matrices $M_\pm$ are computed, the coupled flux for platelets is defined by

$$\mathbf{t}(\phi_c, \phi_f) \approx \mathbf{t}\left(M_+ \begin{pmatrix} \phi_{c,+} \\ \phi_{f,+} \end{pmatrix} + M_- \begin{pmatrix} \phi_{c,-} \\ \phi_{f,-} \end{pmatrix}\right). \tag{5.10}$$

As a result, in the flux vector variation, the matrix coefficients $Q(\Phi)M_+$ have non-negative eigenvalues, whereas $Q(\Phi)M_-$ have non-positive eigenvalues (modified singular values). Both Algorithms 5.1 and 5.2 realize fixed-point iterations that converge either to the boundary of the admissible set of ϕ_c, ϕ_f or to extremum points of $\|\mathbf{t}(\Phi)\|$. The nonlinear advection flux vector discretization mimics the properties of (5.5) but differs in realization.

Remark 5.1 One can apply a stable approximation for all reactive components constituting the vector of concentrations $\Theta = (P, T, B_\alpha, A, F_g, F, F_p, \phi_c, \phi_f)^T$. Approximation for reactions (5.2) is then $R(\Theta^{n+1}) \approx M_+ R(\Theta^{n+1}) + M_- R(\tilde{\Theta}^{n+1})$, where $\tilde{\Theta}^{n+1} = 2\Theta^n - \Theta^{n-1}$. Here, the matrix coefficients M_+ and M_- with property $M_+ + M_- = I$ have order 9. They are obtained from the modified SVD of the Jacobian $Q(\Theta) = \partial R(\Theta)/\partial\Theta = U\hat{\Sigma}\hat{V}^T$, $M_\pm = \frac{1}{2}U\left(\mathrm{sgn}(\hat{\Sigma}) \pm \left|\mathrm{sgn}(\hat{\Sigma})\right|\right)U^T$. The Jacobian is known analytically,

$$
Q(\Theta) = \begin{pmatrix}
\substack{-k_1\phi_c - k_5T^3 \\ -k_3T - k_2B_a \\ -k_4T^2} & P\begin{pmatrix} -3k_5T^2 \\ -k_3 - 2k_4T \end{pmatrix} & -k_2P & & & & & & -k_1P \\
\substack{k_1\phi_c + k_5T^3 \\ +k_3T + k_2B_a \\ +k_4T^2} & \substack{P\begin{pmatrix} 3k_5T^2 \\ +k_3 + 2k_4T \end{pmatrix} \\ -k_6A} & k_2P & -k_6T & & & & & k_1P \\
 & k_8\left(B_0 - B_a\right) & \substack{-k_7\phi_c \\ -k_9A \\ -k_8T} & -k_9B_a & & & & & k_7\left(B_0 - B_a\right) \\
 & -k_6A & -k_9A & \substack{-k_9B_a \\ -k_6T} & & & & & \\
 & -\frac{k_{10}F_g}{K_{10}+F_g} & & & \substack{\frac{k_{10}F_gT}{(K_{10}+F_g)^2} \\ -\frac{k_{10}T}{K_{10}+F_g}} & & & & \\
 & \frac{k_{10}F_g}{K_{10}+F_g} & & & \substack{\frac{k_{10}T}{K_{10}+F_g} \\ -\frac{k_{10}F_gT}{(K_{10}+F_g)^2}} & -k_{11} & & & \\
 & & & & & k_{11} & & & \\
 & -k_{12}\phi_f & & & & & & \substack{-k_{13}\phi_c \\ -k_{12}T} & -k_{13}\phi_f \\
 & k_{12}\phi_f & & & & & & \substack{k_{13}\phi_c \\ +k_{12}T} & k_{13}\phi_f
\end{pmatrix}. \tag{5.11}
$$

The implicit part $M_+R(\Theta^{n+1})$ of the residual $R(\Theta^{n+1})$ contributes to the diagonal block of the Jacobian with eigen-positive matrix coefficient $M_+Q(\Theta^{n+1})$. The explicit part $M_-R(\tilde{\Theta}^{n+1})$ contributes to the right-hand side with eigen-positive matrix coefficient $-M_-$. We avoid iterations of Algorithm 5.2 for computation of $M_\pm$ due to linearization $R(M_+\Theta^{n+1} + M_-\hat{\Theta}^{n+1}) \approx M_+R(\Theta^{n+1}) + M_-R(\tilde{\Theta}^{n+1})$. The SVD decomposition of 9×9 matrix $Q(\Theta^{n+1})$ is required on each nonlinear iteration for each computational cell. The Jacobian matrix is too large for the direct symbolic decomposition; it requires an iterative SVD at each cell $\omega \in \Omega_h$. Simulation time of such fully implicit scheme is greater than the simulation time of the semi-implicit approach presented above.

5.3 Numerical Examples

Coagulation of blood flow is the multi-physics model which involves complex hydrodynamic and reactive processes. In this section, we first demonstrate the efficiency of the FV discretization on general meshes for the Navier–Stokes equations (for details, we refer to [142]), and then we validate the full FV model of clotting blood flow against a real-life experiment.

5.3.1 Lid-Driven Cavity Flow

The lid-driven cavity problem is the benchmark for qualitative reproduction of very high Reynolds number flows. The number and shape of vortices which appear in the flow domain as well as velocity profiles along particular lines evaluate accuracy and low numerical dissipation of a numerical scheme under investigation.

We consider a 3D version of the lid-driven cavity problem [71]. In the unit cube $\Omega = (0, 1)^3$, we introduce a polyhedral prismatic grid with single layer of cells in z-direction. The nodes of an initial regular grid (Fig. 5.2, left) are distorted to improve the boundary layers' resolution. The transformation of the original grid node coordinates x to the new coordinate $\tilde{x}$ is given by

$$\tilde{x} = \frac{1}{2} + \left(x - \frac{1}{2}\right)\left(\frac{1}{4} + \frac{49}{400}\right)^{\frac{1}{2}} \left(\left(x - \frac{1}{2}\right)^2 + \frac{49}{400}\right)^{-\frac{1}{2}}. \tag{5.12}$$

Similar transformation is applied to y-coordinates of each grid node.

The regular and the distorted polyhedral grids with 162 elements are displayed in Fig. 5.2. In the numerical test, we use the distorted polyhedral grid with 65920 elements.

We impose in (2.92) the Dirichlet boundary condition with $\mathbf{u}_b = (1, 0, 0)^T$ at the top side of the cube $\Gamma_{Dir} = \partial\Omega \cap \{y = 1\}$, the slip boundary conditions at the front and back sides of the cube $\Gamma_{slip} = \partial\Omega \cap \{z = 0 \cup z = 1\}$ and the no-slip boundary conditions at the left, right and bottom sides $\Gamma_{noslip} = \partial\Omega \setminus (\Gamma_{Dir} \cup \Gamma_{slip})$. The Reynolds numbers of interest are $Re \in \{100, 400, 1000, 10000, 16000\}$; the viscosity is $\mu = 1/Re$. The free scaling parameter in (2.101) is $\alpha = 1/10$.

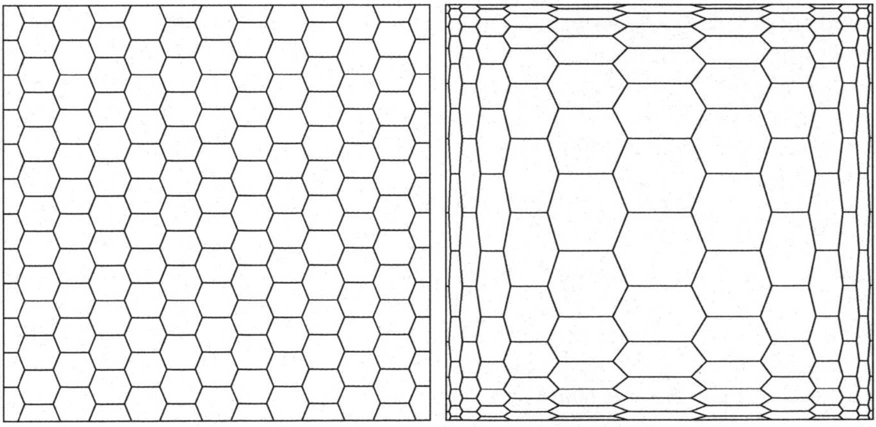

Fig. 5.2 Regular polyhedral grid with 162 elements (left), and grid after distortion (5.12) (right)

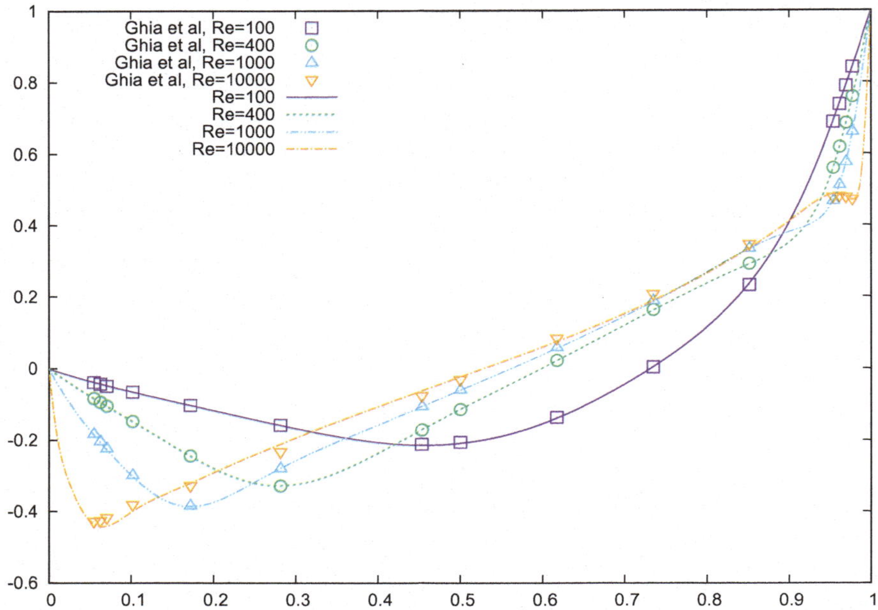

Fig. 5.3 Comparison of the computed u-component of the velocity for $Re = \{100, 400, 1000, 10000\}$ along the line $x = 0.5$, $y = 0.5$ with respect to reference data from [71]

We solve subsequently the steady-state problems for $Re_i \in \{100, 400, 850, 1000\}$, $i = 1, \ldots, 4$. The initial guess to the first problem with $Re_1 = 100$ is zero velocity and zero pressure. We take the steady-state solution of the $(i-1)$th problem as the initial guess to the iterative solution of the ith problem. The iterative solution of the steady-state problems requires from 4 to 7 Newton iterations to reduce the nonlinear residual norm $\|\mathcal{R}\|$ below 10^{-9}. For $Re_4 = 1000$, the Newton iterations do not converge from the initial zero velocity and zero pressure. This is the reason for using the solution of the $(i-1)$th problem as the initial guess to the iterative solution of the ith problem.

For $Re_i \in \{10000, 16000\}$, the unsteady problems (2.91) are solved for $t \in [0, 500]$ starting from zero velocity and zero pressure. The time step $\Delta t_i = t^{i+1} - t^i$ is computed according to the rule: $\Delta t_i = 0.01$, $i = 0, 1, 2$ and $\Delta t_i = \min(\theta \Delta t_{i-1}, \Delta t_{max})$, $i = 3, 4, \ldots, n$, where $\theta = 1.25$, $\Delta t_{max} = 2.5$, $n = \lceil \log_\theta (\Delta t_{max}/\Delta t_0) + 2 \rceil$. At each time step, up to 6 Newton iterations (2.108) are required to reduce the nonlinear residual norm $\|\mathcal{R}\|$ below 10^{-6}. At final time steps t^k, the nonlinear residual norm is below 2×10^{-7} for the solution at t^{k-1}; yet, we perform one Newton's iteration in this case as well. The change in kinetic energy at last steps is of the order of $10^{-4}\%$.

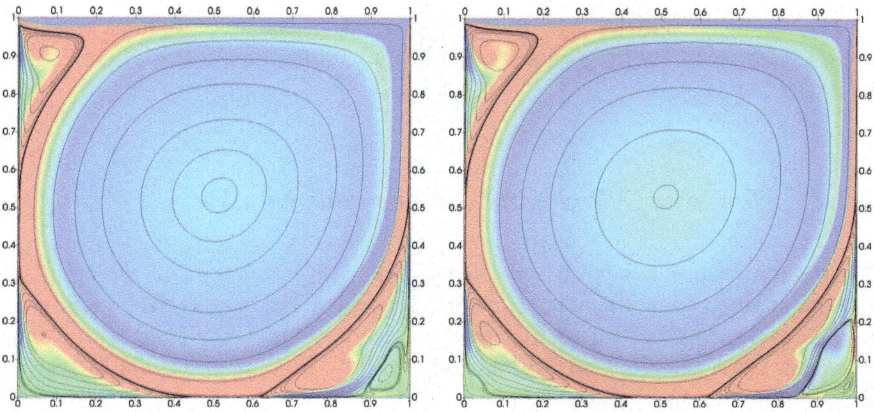

Fig. 5.4 Isolines of stream function ψ for the solution at $Re = 10000$ (left) and $Re = 16000$ (right). The color shows vorticity $\omega_z \in [-2.5, 2.5]$, blue color corresponds to $\omega_z \leq -2.5$, red color corresponds to $\omega_z \geq 2.5$

Figure 5.3 presents the comparison of the computed u-component of the velocity along the line $x = 0.5$, $z = 0.5$ with the reference data from [71] for Reynolds numbers $Re = \{100, 400, 1000, 10000\}$. Very good correspondence of the solutions is observed for $Re = \{100, 400, 1000\}$. For $Re = 10000$, however, the magnitude of the negative extremum is slightly higher and its position is shifted to the right in case of the FV method.

In Fig. 5.4, isolines of the recovered stream function ψ are displayed for $Re = 10000$ and $Re = 16000$. The function ψ is reconstructed based on z-component of the vorticity $\boldsymbol{\omega}$ through the solution of $-\mathrm{div}\nabla\psi = \boldsymbol{\omega}_z$ with Neumann boundary conditions $\nabla\psi \cdot \mathbf{n} = 0$ on front and back sides of the cube and Dirichlet boundary conditions $\psi = 0$ on the remaining boundary. The vorticity $\boldsymbol{\omega}$ is approximated at center $\mathbf{x}_\omega$ of each cell $\omega \in \Omega_h$ by the surface–volume integral formula:

$$\boldsymbol{\omega}|_{\mathbf{x}_\omega} = \frac{1}{|\omega|}\int\limits_{\omega} \nabla \times \mathbf{u}\, d\omega = \frac{1}{|\omega|}\sum_{f\in\mathcal{F}(\omega)}\int\limits_{f} \mathbf{ds} \times \mathbf{u} \approx \frac{1}{|\omega|}\sum_{f\in\mathcal{F}(\omega)} |f|\,\mathbf{n}_f \times \mathbf{u}_f, \tag{5.13}$$

where $\mathbf{u}_f$ is computed according to (2.98) and (2.104) on internal and boundary faces f, respectively. According to Fig. 5.4, reduction of the viscosity from $Re = 10000$ to $Re = 16000$ increases the number of observed vortices: the fourth vortex in the lower right corner and the third vortex in the lower left corner are clearly observed in Fig. 5.4 (right). A precursor to the third vortex in the upper left corner is visible. In a zoom of the velocity vector field, the fifth, the forth, and the third vortex are detectable in the lower right, lower left, and upper right corners of the cavity, respectively. Comparison of the obtained vortices to the reference vortices [30] gives stronger the third vortex of the reference solution in the upper left corner and stronger the fourth vortex of the examined FV solution in lower right corner.

The presented velocity and vorticity streamlines indicate to high accuracy and low dissipation of the FV discretization (2.93) with the flux vector discretizations (2.100), (2.105).

5.3.2 Flow over Cylinder at Low Reynolds Number

The benchmark problem [130] is devoted to quantitative validation of incompressible flow simulators at low Reynolds numbers. The benchmark targets the pressure drop as well as the drag and lift coefficients for the flow around a cylinder with circular or square cross section. Further deliberation of the benchmark results for the low Reynolds number $Re = 20$ is available in [40]. We shall compare the statistics of the FV solution (2.93), (2.100), (2.105) with the reference statistics from both sources.

The computational domain is $\Omega = [0, L] \times [0, H] \times [0, H] \setminus \Omega_X(\mathbf{x}_0, \mathbf{x}_1, D)$, where $\Omega_X(\mathbf{x}_0, \mathbf{x}_1, D)$ defines the cylinder with circular ($\Omega_X = \Omega_C$) or square ($\Omega_X = \Omega_S$) cross section of diameter D with the centers of bases at $\mathbf{x}_0 = (0.5, 0.2, 0)^T$ and $\mathbf{x}_1 = (0.5, 0.2, H)^T$. The parameters of the domain are $L = 2.5$, $H = 0.41$, $D = 0.1$. Note that the cylinder centerline does not belong to the symmetry plane of Ω, and thus the lift coefficient is non-zero.

We consider two classes of meshes, tetrahedral and polyhedral ones. Each polyhedral mesh is constructed by approximate dualization of corresponding tetrahedral meshes. In case of the circular cylinder, after dualization of the mesh, its boundary nodes approximating the cylinder surface are projected onto the surface. We exploit four Delone tetrahedral meshes with indices 1, 2, 3, and 4. The initial mesh 1 is constructed using GMSH [70], and subsequent meshes are refined by splitting and optimized using Netgen [132]. By indices $a^\dagger$, we denote a locally adapted mesh constructed with GMSH and Netgen with complex refinement pattern to cylinder Γ_X, to boundaries and inside of the domain. The cutaway of the dualized mesh $a^\dagger$ is displayed in Fig. 5.5.

We impose in (2.92) the Dirichlet boundary condition with

$$\mathbf{u}_b = \left(36\hat{U}yz(H-y)(H-z)/H^4, 0, 0\right)^T,$$

at the inflow $\Gamma_{Dir} = \partial\Omega \cap \{x = 0\}$, the do-nothing boundary conditions at the outlet $\Gamma_{do-noth} = \partial\Omega \cap \{x = L\}$, and the no-slip boundary conditions at the remaining boundary $\Gamma_{noslip} = \partial\Omega \setminus (\Gamma_{Dir} \cup \Gamma_{do-noth})$. According to the benchmark, fluid viscosity is $\mu = 0.001$ and density is $\rho = 1$. Independent of time characteristic velocity $\hat{U} = 0.2$ defines the Reynolds number $Re = 20$ and produces a steady-state solution of (2.91).

The drag and lift coefficients are defined by the volumetric integrals over the domain Ω [40, 120]:

$$
\begin{aligned}
C_D &= -C \int\limits_{\Gamma_X} \left(p\mathbf{n}_x - \mu \frac{\partial v_\tau}{\partial \mathbf{n}} \mathbf{n}_y \right) \mathrm{d}s \\
&\approx C \sum_{\omega \in \mathcal{V}(\Omega_h)} |\omega| \left(p \frac{\partial \phi}{\partial x} - (\rho \phi \mathbf{u} + \mu \nabla \phi) \cdot \nabla u \right), \\
C_L &= -C \int\limits_{\Gamma_X} \left(p\mathbf{n}_y + \mu \frac{\partial v_\tau}{\partial \mathbf{n}} \mathbf{n}_x \right) \mathrm{d}s \\
&\approx C \sum_{\omega \in \mathcal{V}(\Omega_h)} |\omega| \left(p \frac{\partial \phi}{\partial y} - (\rho \phi \mathbf{u} + \mu \nabla \phi) \cdot \nabla v \right),
\end{aligned}
\tag{5.14}
$$

where $C = 2/(\rho \hat{U}^2 DH)$, v_τ is the tangential velocity along the tangent $\tau = (\mathbf{n}_y, -\mathbf{n}_x, 0)^T$ to Γ_X, $\mathbf{n}$ is the normal to face f oriented inside Ω. Function ϕ is obtained from solution of div$\nabla = 0$ with Dirichlet boundary conditions $\phi|_{\Gamma_X} = 1$ and $\phi|_{\partial\Omega \setminus \Gamma_X} = 0$. The values of pressure p and gradients of velocity components ∇u and ∇v for (5.14) are taken from cell ω.

Another quantity of interest is the pressure drop $\Delta p = p(\mathbf{x}_a) - p(\mathbf{x}_e)$, where $\mathbf{x}_a = (0.45, 0.2, 0.205)^T$ and $\mathbf{x}_e = (0.55, 0.2, 0.205)^T$. The pressure drop is approximated by taking pressure values p_a and p_e from cells ω_a and ω_e with centers closest to $\mathbf{x}_a$ and $\mathbf{x}_e$, respectively.

The steady solution of the steady counterpart of (2.91) is found by the Newton iterations with the nonlinear tolerances $\tau_{abs} = 10^{-6}$, $\tau_{rel} = 10^{-7}$. It takes 4–5 iterations to meet the convergence criteria. The free parameter α is equal to $1/12$. The computed statistics are presented in Table 5.1. The polyhedral adapted mesh a†, and the corresponding FV solution of the problem with the circular cylinder are depicted are Fig. 5.5.

The drag and lift coefficients and the pressure drop of the FV solution are close to the reference values. Aggressive refinement in a proximity of the circular cylinder is required to match the reference data. All the coefficients are resolved slightly better with tetrahedral grids which, however, result in considerably more degrees of freedom.

The presented statistics confirm the high accuracy and the low dissipation of the FV discretization (2.93) with the flux vector discretizations (2.100), (2.105).

5.3.3 Coagulation of Blood Flow in Microfluidic Capillaries

The multi-physics model of blood flow coagulation (5.1)–(5.2) is validated across in vitro experiment in microfluidic capillaries [134]. In this experiment, the flow of platelet-rich blood plasma (PRP) is driven by a constant pressure drop in a microfluidic capillary. The characteristic sizes of the capillary are given in Fig. 5.6. The damage to the blood vessel causes the inflow of the tissue factor (TF) into the blood-

Table 5.1 Estimated lift and drag coefficients and pressure drop for flows around cylinders with circular and square cross section

Level	Circular cross section (3D-1Z)				Square cross section (3D-1Q)			
	$\|\mathcal{V}(\Omega)\|$	C_D	C_L	Δp	$\|\mathcal{V}(\Omega)\|$	C_D	C_L	Δp
Grid type	Tetrahedral grid							
1	2609	4.261	−0.04722	0.1060	3987	6.985	0.15697	0.1423
2	18749	5.124	0.07479	0.1451	29184	7.544	0.04563	0.1630
3	139643	6.070	0.01267	0.1612	218486	7.626	0.07611	0.1698
4	1052665	6.148	0.00985	0.1634	1654601	7.627	0.07102	0.1713
a[a]	279608	6.151	0.01022	0.1665	300287	7.720	0.06347	0.1737
Grid type	Polyhedral grid							
1	729	5.176	0.05416	0.1025	967	7.362	0.17674	0.1604
2	4224	6.296	0.05653	0.1665	6002	8.459	0.10862	0.1839
3	27553	6.498	0.00909	0.1780	40840	8.171	0.10128	0.1801
4	192361	6.219	0.00725	0.1678	293283	7.753	0.08423	0.1723
a[a]	58982	6.221	0.00964	0.1641	63632	7.858	0.05974	0.1743
Schäfer & Turek [130]	–	6.05–6.25	0.008–0.01	0.165-0.175	–	7.5–7.7	0.06-0.08	0.172–0.18
Braack & Richter [40]	–	6.185331	0.00940136	0.171342	–	7.76[b]	0.0688	0.17567

[a]Mesh with local refinement
[b]Note that the value provided by Braack & Richter is outside of the range provided by Schäfer & Turek

stream. The TF in the blood flow initiates reactions that cause blood coagulation. A TF-patch on the interior surface of the tube depicted red in Fig. 5.6 is the source of the tissue factor.

The experiment [134] studies the initiation time of blood coagulation and the time of complete vessel occlusion as functions of the shear rate γ parameter of the flow.

For 0.2 mm width of the TF-patch, the in vitro experiment demonstrates the following. When the shear rate is lower than the threshold value $30\,\mathrm{s}^{-1}$, coagulation is initiated at $t = 100\,\mathrm{s}{-}200\,\mathrm{s}$, the flow is completely stopped, and the stagnant plasma turns into a thrombus, which completely occludes the vessel. When the value of the shear rate exceeds $30\,\mathrm{s}^{-1}$, the initiation of the coagulation process happens later than $t = 800\,\mathrm{s}$. Under the fixed shear rate $40\,\mathrm{s}^{-1}$, coagulation is initiated at $t = 100\,\mathrm{s}{-}200\,\mathrm{s}$ for the width of the TF-patch higher than 0.4 mm, and later than $t = 800\,\mathrm{s}$ for the width smaller 0.4 mm.

The flow of PRP in microcapillaries may be considered as flow of Newtonian fluid; therefore, the multi-physics model from Sect. 5.1 is applicable to the above experiment. The flow is driven by the pressure drop Δp condition in (2.92) which defines the flow shear rate according to the Pousielle solution. The computational mesh is shown in Fig. 5.6.

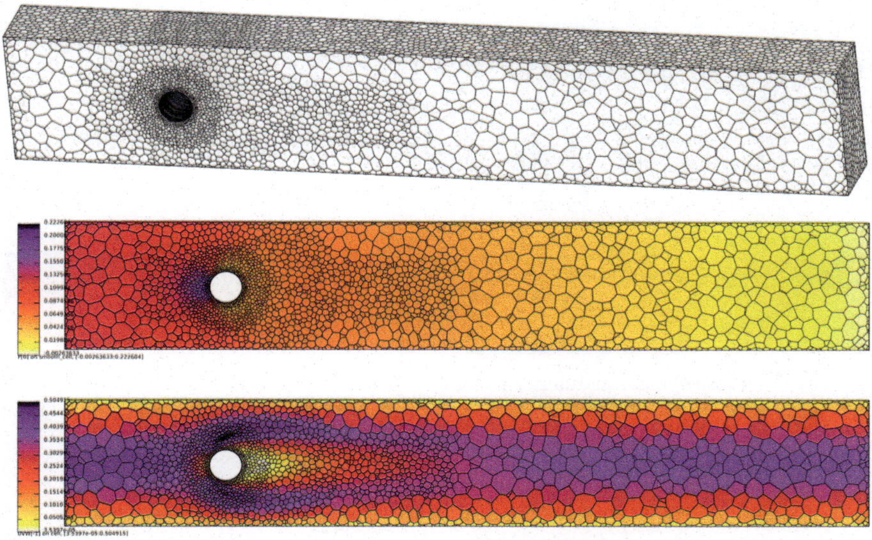

Fig. 5.5 A middle cutaway $z = 0.205$ of the polyhedral adaptive mesh a† constructed for the domain with circular cylinder (top), colored with the FV pressure field (middle) and FV velocity magnitude (bottom)

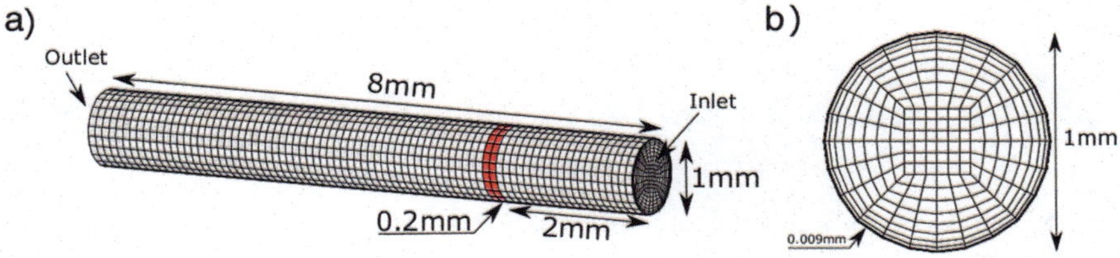

Fig. 5.6 **a)** Dimension of computational domain and TF-patch location on the grid. **b)** Cross section of the grid demonstrating local refinement to cylinder surface

Figures 5.7 and 5.8 represent the FV solution of (5.1)–(5.2) in the PRP at the shear rate $\gamma = 25\,\mathrm{s}^{-1}$ and times $t = 60\,\mathrm{s}$ and $t = 70\,\mathrm{s}$, respectively. For illustration purposes, the velocity magnitude is demonstrated in the log scale. From the velocity distribution, it is evident that the flow is obstructed by the clot at time $t = 70\,\mathrm{s}$. The numerical experiment recovering shear rate $\gamma = 25\,\mathrm{s}^{-1}$ reproduces the basic characteristics of the in vitro experiment. For complete comparison and details on the initial and boundary conditions, as well as the model parameters, we refer to [39].

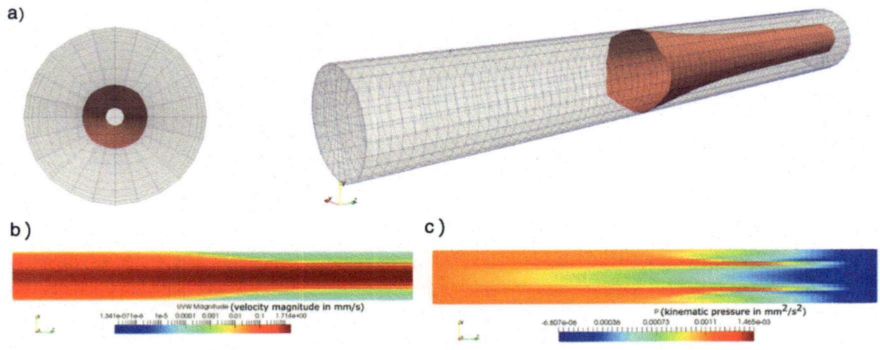

Fig. 5.7 Simulation of blood clotting at shear rate $\gamma = 25\,\mathrm{s}^{-1}$, moment of time $t = 60\,\mathrm{s}$. **a)** Isosurface corresponding to permeability coefficient $K_f = 1$. **b)** Log scale of velocity field in a middle cutaway of grid. **c)** Pressure field in a middle cutaway of grid

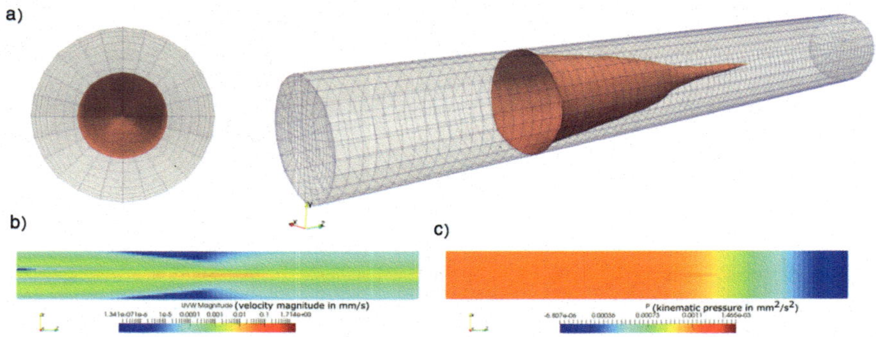

Fig. 5.8 Simulation of blood clotting at shear rate $\gamma = 25\,\mathrm{s}^{-1}$, moment of time $t = 70\,\mathrm{s}$. **a)** Isosurface corresponding to permeability coefficient $K_f = 1$. **b)** Log scale of velocity field in a middle cutaway of grid. **c)** Pressure field in a middle cutaway of grid

Table 5.2 Performance of the multi-physics model of blood flow coagulation on different numbers of processors

Processors	Assembly time		Solution time		Total time		Equations per proc.
1	2781	–	2156	–	4938	–	318500
8	461	6x	214	10x	680	7x	39812
16	253	11x	117	18x	373	13x	19906
32	148	19x	80	27x	234	21x	9954
64	90	31x	46	47x	141	35x	4977
128	46	60x	29	74x	77	64x	2489

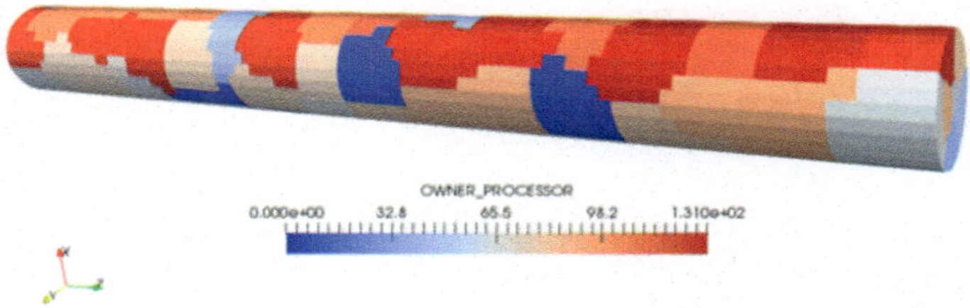

Fig. 5.9 Partitioning of the computational mesh among processors. The mesh represents the cylindric vessel corresponding to experiment [134]

5.3.3.1 Parallel Performance

The parallel performance of the multi-physics model of blood flow coagulation is demonstrated in Table 5.2. The model scales very well with the growth of the number of processors. The reason for the good scalability is that the main contribution to the arithmetical work is given by the reactions term which is computed on each processor independently. The partitioned mesh is displayed in Fig. 5.9.

Chapter 6
INMOST Platform Technologies for Numerical Model Development

Integrated Numerical Modeling Object-oriented Supercomputing Technologies (INMOST) is an open-source, flexible, and efficient numerical modeling framework that provides to application developers all the tools required for fast development of parallel multi-physics models. The users of INMOST avoid designing and implementing their own mesh data structures. They enjoy low-level infrastructure for reading, writing, creating, manipulating, and partitioning distributed unstructured general meshes.

FV discretizations of systems of PDEs may lead to systems of nonlinear algebraic equations. The nonlinear algebraic equations are solved iteratively. On each nonlinear iteration, a sparse linear system corresponding to the nonlinear residual is assembled and solved. INMOST provides software tools that cover the complete process for parallel assembly and solution of linear systems arising in the iterative solution of nonlinear systems. To simplify the solution, the user can use INMOST tool for automatic differentiation for assembly of the nonlinear residual and the corresponding Jacobian and Hessian matrices.

The synergy of the monotone FV discretizations for systems of PDEs on general meshes and INMOST instruments for development of numerical models on general meshes produces a powerful tool for supercomputing simulations. This chapter describes INMOST functionality and software mechanisms for interacting with it.

6.1 Maintenance of General Meshes

INMOST mesh maintenance exploits full mesh representation with circular adjacencies node–edge–face–cell–node which provides the balance between memory requirements and parallel algorithms efficiency. The following mesh elements are supported:

Y. Vassilevski et al., *Parallel Finite Volume Computation on General Meshes*,
https://doi.org/10.1007/978-3-030-47232-0_6

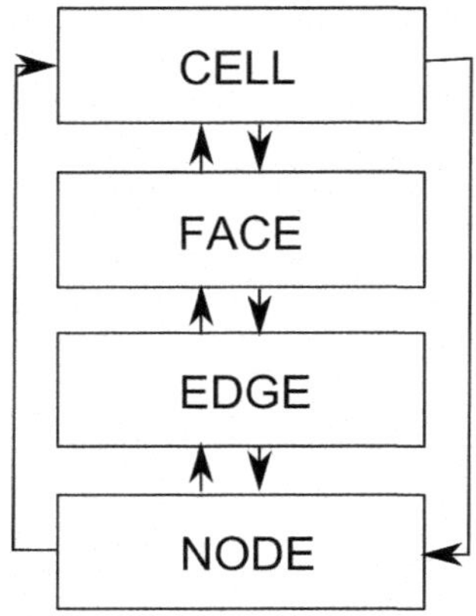

Fig. 6.1 Hierarchy of mesh elements

- Node, which contains information on the position in the space;
- Edge, which consists of two or more vertices;
- Face (polygon in general case), which is based on a set of edges;
- Cell (polyhedron in general case), which is based on a set of faces.

Mesh elements can be specified by their dimension and geometric type:

- (0D) Node: Vertex;
- (1D) Edge: Line;
- (2D) Face: Triangle, Quad, Polygon;
- (3D) Cell: Tetrahedron, Hexahedron, Prism, Pyramid, Polyhedron.

These mesh elements are organized in the following hierarchy (Fig. 6.1): Cell $\leftrightarrow$ Face $\leftrightarrow$ Edge $\leftrightarrow$ Node. Nodes, edges, faces, and cells form adjacency connections between them: an edge is based on nodes, a face is based on edges, and a cell is based on faces or on nodes. The traversal of adjacencies should be ordered: edges of each face form a loop that defines the normal orientation, and nodes of certain types of cells appear in a predefined order. INMOST can support the abovementioned connections as well as a complete set of element connections at the cost of additional memory. Adjacency connections facilitate search of neighboring elements that is important for implementation of FV discretizations.

INMOST platform supports the concept of mesh element data:

- *Elements* are used to store a mesh configuration;
- *Data* are used to store information in mesh elements;
- *Tags* are used to connect mesh data to mesh elements;
- *Sets* are used to organize mesh *Elements*.

Mesh functions of INMOST operate with *Data*, *Tags*, *Elements*, *Sets of elements*, and *Mesh*. Mesh *Elements* can be organized into *Sets of elements* which are useful in assigning boundary conditions and other operations associated with certain elements. Mesh includes all *Tags* and all *Elements* of the mesh. *Sets of elements* may be

Fig. 6.2 Organization of sets of elements into a tree structure

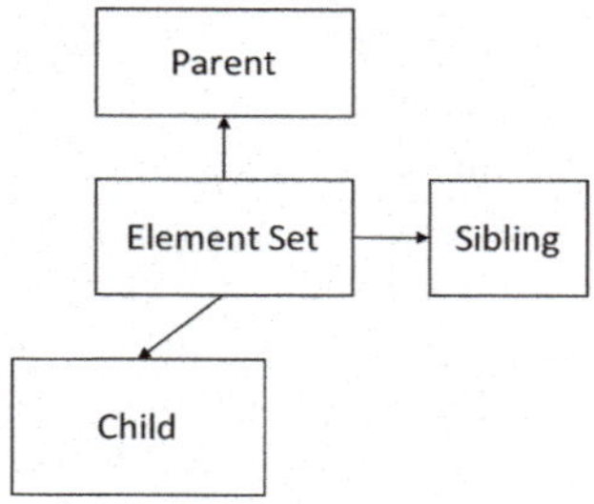

organized into a hierarchical tree structure (Fig. 6.2). Such structure is useful in many cases, i.e., for mesh adaptation and for representation of groups of wells connected to the mesh.

Data is associated to a mesh element by a *Tag*. Mesh data may be represented in a variety of ways (Fig. 6.3). Dense data are associated with all elements, and sparse data are given on some elements. Data can have fixed or variable size, as well as various data types: bulk (single character), integer, double, a reference to an element, double with single or multiple variations, and a reference to an element of another mesh. Variation is represented by a sparse vector consisting of indices and coefficients that represent partial derivatives. This type of data is used to store intermediate results of Jacobian's calculation. Data are accessible directly in the memory through provided classes or may be copied to provided arrays.

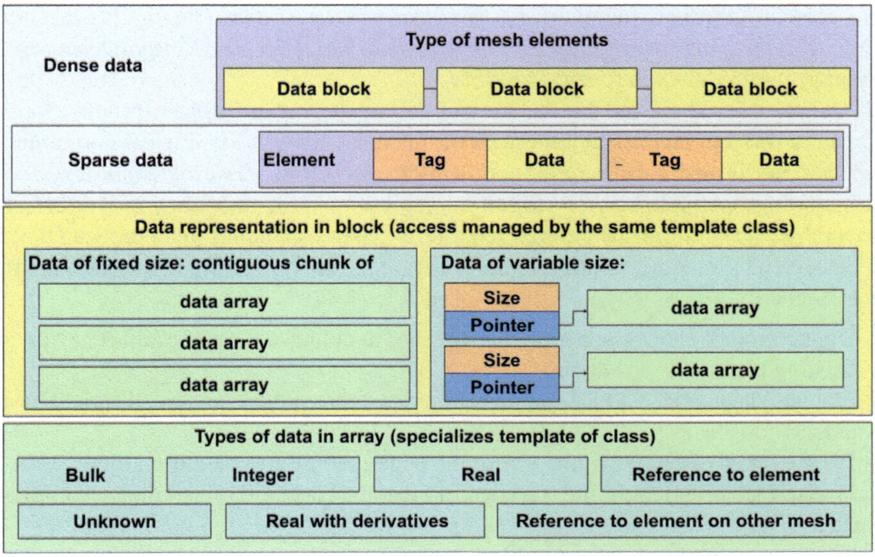

Fig. 6.3 Data representation in a mesh

6.2 Generation and Modification of General Meshes

Any mesh is assumed to be consistent in a sense that any two cells may share one entire face. Mesh generation consists of addition of new elements and deletion of existing elements. Addition or deletion of an element is performed by modification of its adjacency connections.

Deletion of an element implies disconnection of lower level adjacency connections[1] and deletion of all adjacency elements dependent on the deleted element. These operations guarantee consistency of the mesh. Functions *Element::Connect, Element::Disconnect* connect and disconnect adjacencies of mesh elements. Function *Element::Delete* deletes or hides an element (depending on a mesh state), whereas function *Element::Destroy* destroys an element unconditionally.

Insertion of a new element into the mesh is performed by introduction of its lower adjacency dependence. Functions *Mesh::CreateNode*, *Mesh::CreateEdge*, *Mesh::CreateFace*, *Mesh::CreateCell* create an appropriate mesh element.

Modification of a mesh element is a two-stage procedure. First, upper adjacencies of the element are disconnected and the element is deleted. This stage makes the mesh inconsistent. Second, new elements are added to the mesh, and the upper adjacencies are reconnected by function *Element::Connect*. This makes the mesh consistent again.

Useful geometrical quantities (edge length, face area and normal, cell volume, element barycenter) can be computed and stored by function *Mesh::PrepareGeometricData*. Function *Mesh::RemoveGeometricData* removes geometrical data. Function *Mesh::HaveGeometricData* checks data availability. The geometrical data are accessed through functions *Element::Barycenter*, *Edge::Length*, *Face::Area*, *Cell::Volume*, *Face::Normal*, *Face:: OrientedNormal*, *Face::FixNormalOrientation* or, in general, *Mesh::GetGeometricData*.

Deletion, insertion, and modification of a mesh element are prone to produce topological errors (an inconsistent mesh). For this reason, mesh topological correctness tests are performed during mesh modifications. Function *Mesh::SetTopologyCheck* sets topology test, function *RemTopologyCheck* removes it, and function *Mesh::GetTopologyCheck* returns the current set of tests. Wrong elements get a tag which can be accessed by function *Mesh::TopologyErrorTag*. The following topological errors are checked:

- Duplication of elements: a new element should not have the same adjacency as an existing element.
- Element degeneracy: a face should have at least three edges, and a cell should have at least four faces.
- Wrong order of edges within a face: the edges should form a closed loop.
- Wrong face orientation: the traversal of face nodes should match the face normal direction.
- Face non-planarity: a face should be planar.

[1]Two nodes for an edge, edges, or nodes for a face, faces, or nodes (in some cases) for a cell.

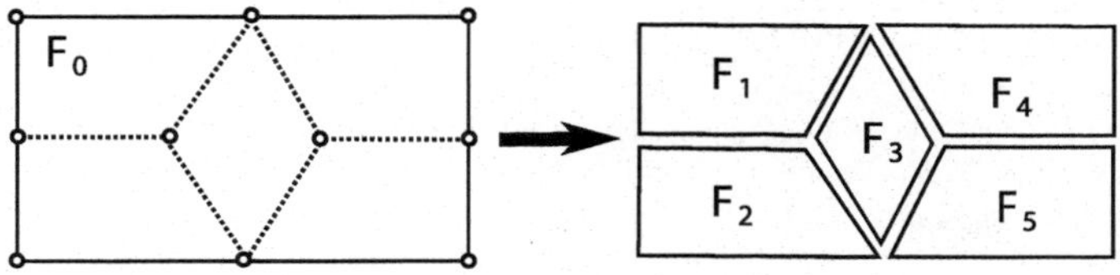

Fig. 6.4 Splitting of face F_0 by a set of edges into faces $F_1 ... F_5$

- Interleaved faces: no multiple faces may share the same nodes.
- Mesh inconsistency: each interior face should be shared by exactly two neighboring cells.
- Slivers: a cell face should not contain all the cell nodes.
- Duplication of adjacencies, disconnected adjacencies, improper dimensionality of adjacencies.
- Lower adjacencies of elements do not form a closed loop.
- Unknown types of elements.

Local mesh modification implies establishing all necessary reconnections in the mesh. Two types of high-level mesh modification routines are provided by INMOST: uniting a set of elements into a single element and splitting an element into subelements.

Under unification, all lower level adjacencies internal to the union are detected and eliminated, the new element is connected to lower level adjacencies external to the union, and upper level adjacencies of the new element are reconnected. Note that merging elements is not always possible without topological errors: if a hexahedron is surrounded by other hexahedra, its faces cannot be united. Functions *Edge::UniteEdges*, *Face::UniteFaces*, and *Cell:UniteCells* unite sets of edges, faces, and cells, respectively.

For splitting, functions *Edge::SplitEdge*, *Face::SplitFace*, and *Cell::SplitCell* split an edge by nodes, a face by edges, and a cell by faces. Dividing a face by a set of edges and a cell by a set of faces requires to find in the graph formed by the adjacency connections, all closed loops with minimal geometrical measure (length, area, or volume). The search of these loops is performed by a recursive algorithm implemented in functions *Face::SplitFaces*, *Cell::SplitCell*. An example of dividing a face by a set of edges is shown in Fig. 6.4.

Mesh modification is often accompanied by transfer of data attached to the mesh. Transfer of physical quantities is not performed during mesh modification since data interpolation depends on the problem and the data. The user can implement interpolation procedures on the basis of the concept of modification epochs. To reduce the memory footprint, we do not require to store the original mesh before modification. The mesh may exist in a special modification state, when deleted elements are hidden from the mesh, but their data are still available for data transfer. These data can be used for data interpolation, see Algorithm 6.1:

Algorithm 6.1 Mesh modification epoch

1: Call *Mesh::BeginModification* to enter the modification state.
2: Perform modification of the mesh by deleting old elements and creating new elements.
3: Call *Mesh::ResolveModification* to setup the data necessary for data transfer on the new mesh.
4: Call *Mesh::SwapModification* to recover the old mesh and transfer mesh data.
5: Call *Mesh::ApplyModification* to apply all changes. All the elements marked for deletion and their data are irreversibly destroyed.
6: Call *Mesh::EndModification* to exit the modification state.

The tools presented above allow the user to perform mesh generation, mesh repair, mesh improvement, and mesh adaptation. We demonstrate applications of the tools for examples which are available in the INMOST repository [7].

Mesh inconsistency. The *Corner Point Grid* format stores a geological grid with a fault in a simple way: it is a combination of two grids which are shifted vertically with respect to each other. The two grids are disjoint, and thus the geological grid is not a consistent mesh. An example of such geological grid with a fault is shown in Fig. 6.5.

To make the grid consistent, *GridTools/FixFaults* finds intersection of contacting faces, inserts new edges, and splits the contacting faces as depicted in Fig. 6.6.

Fig. 6.5 A mesh with fault

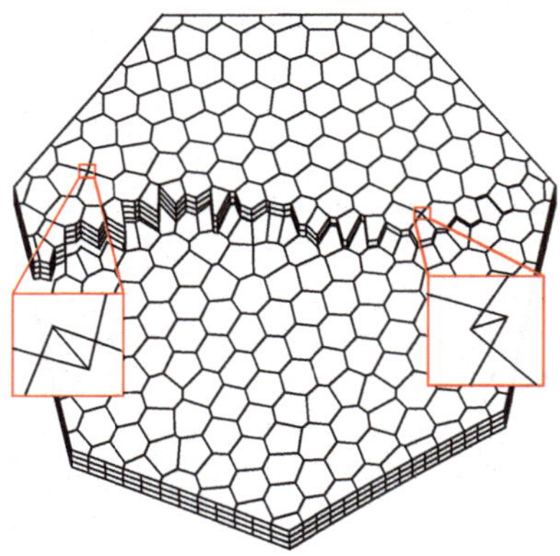

Fig. 6.6 Intersection of contacting faces and splitting them into a new set of faces

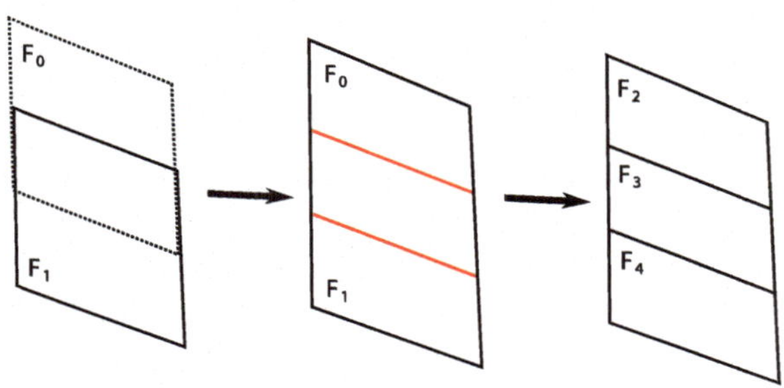

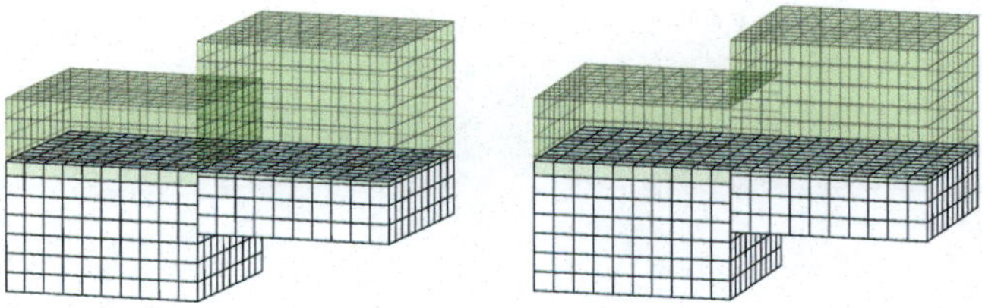

(a) (b)

Fig. 6.7 Traces of two disjoint meshes on a fault (**a**). Faces are split at the fault between two meshes (**b**)

An example of two disjoint mesh traces on a fault is shown in Fig. 6.7a. The corrected consistent grid on the fault is shown in Fig. 6.7b.

Mesh irregularity. In practice, meshes may have faces or edges of very small size which affect the stability and accuracy of discretization methods, result in very stiff matrices, etc. One can collapse such elements: an edge may be collapsed to a node, a face may be collapsed to an edge or a node, a cell may be collapsed to a face, an edge, or a node. The collapse operation for a cell is based on the analysis of the bounding ellipsoid with minimum volume [145]. One measures semi-axes of the ellipsoid and

- collapses the cell to an edge, if one semi-axis is significantly larger than the others;
- collapses the cell to a face, if the two largest semi-axes are close;
- collapses the cell to a node, if all three semi-axes are small and almost equal.

The tool *GridTools/FixTiny* repairs tiny elements in the grid presented in Fig. 6.5.

Cutting grids. The tools *GridTools/Slice*, *GridTools/SliceFunc* cut cells by a plane or by zero-level of a level set function. For instance, *GridTools/Slice* can form geological layers within a uniform grid cutting it by planes, *GridTools/SliceFunc* can produce a mesh in a domain of interest, via cutting a uniform grid by a level set function. Tiny cells produced in both examples can be collapsed by the tools discussed above. An example of cutting a uniform grid by a level set function is demonstrated in Fig. 6.8.

Dual meshes. The tool *GridTools/Dual* builds the dual grid from a given mesh. The dual grid contains less cells than tetrahedral or triangular prismatic grids although the cells in the dual grid have more faces. Dual grids are used extensively in vertex-centered FV methods. An example of dual to triangular prismatic grid is shown in Fig. 6.9.

Mesh adaptation. The tool *GridTools/AdaptiveMesh* provides dynamic mesh adaptation of general polyhedral meshes. Dynamic adaptation implies subsequent local refinement and local coarsening.

Refining of a cell requires the addition of nodes at centers of the cell, its faces, and edges. For each cell edge, a quadrilateral is formed by the mid-edge, the centers of two

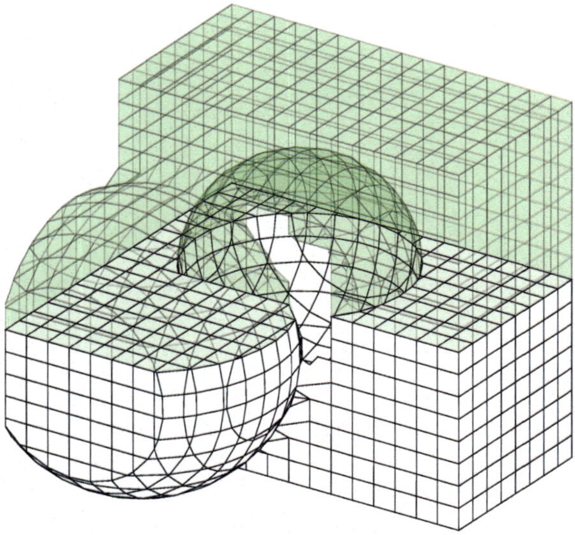

Fig. 6.8 Cutting the grid for a domain which is defined as interior of a sphere in one half of the grid and exterior of another sphere in the second half of the grid. Transparent green surface defines the boundaries of the mesh, and mesh cuts are displayed with the gray color

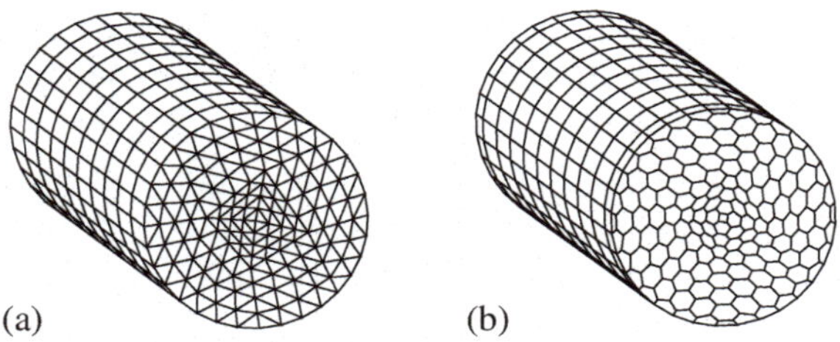

Fig. 6.9 Triangular prismatic grid (**a**), its dual grid (**b**)

cell faces adjacent to the edge, and the cell center (Fig. 6.10). The cell is partitioned into subcells by these quadrilaterals. Recovery of consistency of the refined mesh completes the local refinement.

The following rules for local refinement of a general polyhedral cell are applied:

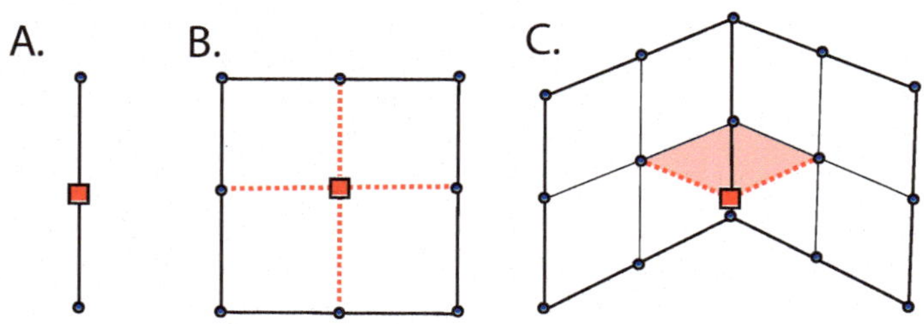

Fig. 6.10 Steps for splitting edge (**a**), face (**b**), and introduction of faces splitting a cell (**c**). Red color indicates new elements on each step

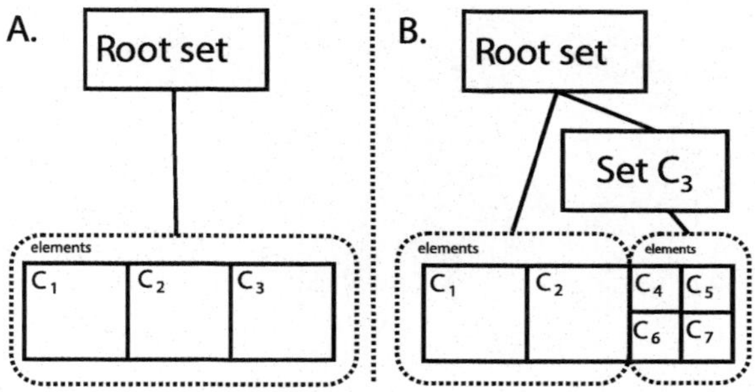

Fig. 6.11 Original unsplit mesh of three cells (**a**). After splitting of one of the cells, a new set is attached to the parent (root) set (**b**). The set remembers all the fine cells that formed original cell

- An edge is split into two edges by a node, added at the mid-edge, Fig. 6.10a. The new node becomes the hanging node of all adjacent faces.
- A face is split only if all its edges are split. A new node is added to the face. Each coordinate of the node is the average of the respective coordinates of face hanging nodes. New edges connecting the latter hanging nodes and the new node are added to split the face into subfaces, as illustrated in Fig. 6.10b.
- A cell is split only if all its faces are split. A new node is added to the cell. Each coordinate of the node is the average of the respective coordinates of cell hanging nodes. New edges connecting the latter hanging nodes and the new node are added to form new faces which split the cell into subcells, as illustrated in Fig. 6.10c. The new faces are not necessarily flat.

Implementation of these rules is based on references to hanging nodes.

Local coarsening is based on hierarchy of sets of mesh elements, see Fig. 6.2. If a cell is refined, a new set (with a unique name) representing that cell is attached to the parent (root in Fig. 6.11) set. All the new cells become the elements of this set as illustrated in Fig. 6.11.

The coarsening of cells can be performed only on the leaf set of the tree structure which was created during refinements. The coarsening for the cells follows these steps:

- Find a node that is shared by all the cells from the leaf set. This node is connected by edges to the future hanging nodes of the coarse cell.
- Unite the cells into a coarse cell.
- Connect the coarse cell to the parent of the leaf set.
- Add the references of all the hanging nodes to the coarse cell.
- If a face has higher level of refinement than both adjacent cells, such face has to be coarsened.
- If an edge has higher level of refinement than all the adjacent faces, we unite the two edges shared by these faces.

On each sweep over the mesh, we do not refine or coarsen by more than one level.

Implemented in INMOST rules of local refinement and coarsening define the robust method of local mesh adaptation. The only requirement for the mesh is its topological consistency and star-shapeness of its elements (cells and faces). The application of the method to the three types of general grids is illustrated in Fig. 6.12.

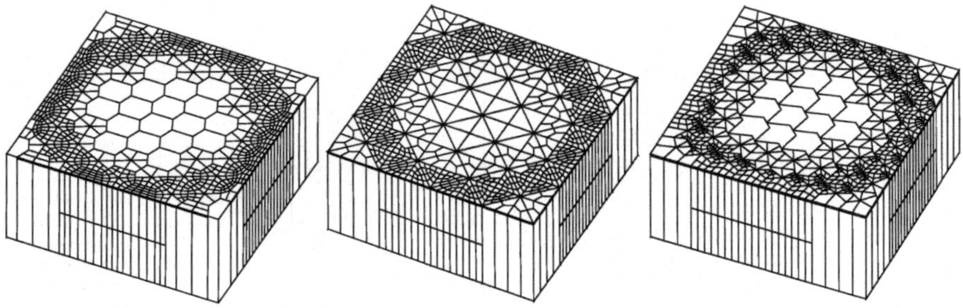

Fig. 6.12 Local adaptation of various prismatic meshes, from left to right: hexagonal, triangular, non-convex squama. The middle cutaway of the grids

6.3 Parallel Mesh Operations

INMOST paradigm of distributed mesh data is based on parallel running on several processors, domain decomposition, and overlapping grids with ghost cells. Every processor stores a part of the whole mesh. Partitioning the mesh into parts is performed by a load balancing algorithm. FV discretizations require a shell of neighboring cells around every cell. In order to provide the shell of neighbors for cells residing near the interface boundary of the local mesh part, we form additional overlapping layers of cells. Different discretization stencils require different types and widths of overlapping layers. These additional cells are exact copies of cells that reside on another processor sharing the interface with the former processor. We call these cells as *ghost* cells. In INMOST, each mesh element is assumed to have exactly one processor–owner. An element that resides on its owner processor and has corresponding ghost copies on other processors is called *shared* element. Each shared element also stores the list of processors it is copied to. The main difference between the ghost element and the normal element is that data in the ghost element should be actualized after update in the normal element. This entails interprocessor communication that moves data from shared cells to their ghost copies.

Further, we will use the following supplementary mesh data tags:

- "State", one of the following states of the element (Fig. 6.13):
  - "owned": the element has no copies on other processors,
  - "shared": the element has copies on other processors,
  - "ghost": the element is a copy.
- "Owner" is the processor that owns the element.
- "Processors" is array of processors that possess the element.

We distinguish two ways for parallel exchange of mesh elements: migration and ghosting. During migration, the original copy of the element is deleted and the element is recreated at a remote processor. The remote processor is assigned as the

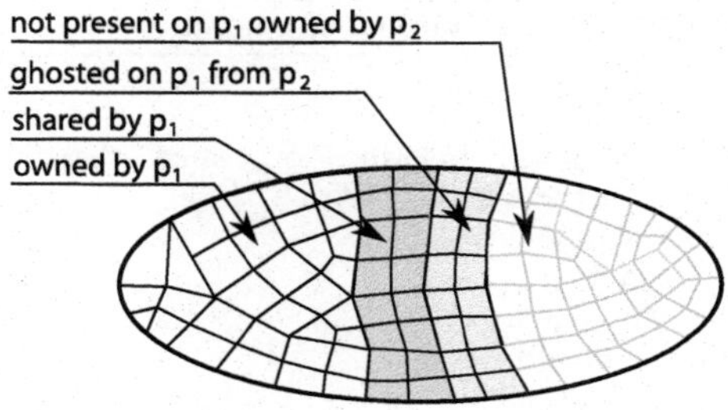

Fig. 6.13 Owned, shared, and ghost elements for a distributed mesh for processor p_1

new owner of the element. During ghosting, a copy of the element is created at a remote processor. The owner of the element remains the same.

Before we present basic parallel algorithms for distributed mesh data related to ghosting and migration, we dwell on a few auxiliary algorithms.

Interprocessor exchange of raw data. Suppose the current processor p exchanges raw data of known length with the sets of processors $\mathcal{P}_{send}, \mathcal{P}_{recv}$. Let $\mathbb{B}_{p,out}$ be the outgoing data array assigned for processor p, $\mathbb{B}_{p,in}$ is the space for incoming data array from processor p, operation size of$(\cdot)$ returns the size of the data array, and $r_p \in \mathcal{R}$ and $s_p \in \mathcal{S}$ represent the handles that help to track the state of the sent data. Algorithm 6.2 represents the simplest blocking version of data exchange that corresponds to function *Mesh::ExchangeBuffersInner*.

Algorithm 6.2 Raw data exchange

1: **for** $p \in \mathcal{P}_{recv}$ **do**
2: MPI_Irecv($\mathbb{B}_{p,in}$, sizeof($\mathbb{B}_{p,in}$), p, r_p)
3: **end for**
4: **for** $p \in \mathcal{P}_{send}$ **do**
5: MPI_Isend($\mathbb{B}_{p,out}$, sizeof($\mathbb{B}_{p,out}$), p, s_p)
6: **end for**
7: **while** MPI_Waitsome($r_k \in \mathcal{R}$) **do**
8: process message corresponding to r_k
9: **end while**
10: MPI_Waitall($\mathcal{S}$)

If we cannot estimate the size of arrays $\mathbb{B}_{p,in}$, we can use the same algorithm to exchange sizes first with sizeof($\mathbb{B}_{p,in}$) $=$ sizeof($\mathbb{B}_{p,out}$) $=$ sizeof(int). If the set $\mathcal{P}_{recv}$ is unknown, then we may use MPI_Allreduce to compute the number of incoming messages by summing up outgoing messages $\mathcal{P}_{send}$ across processors. This functionality is realized in *Mesh::PrepareReceiveInner* with many available strategies using different MPI tools (i.e., MPI-2 point-to-point communications).

Global elements enumeration. Enumeration Algorithm 6.3 prescribes unique identificator to every element $e \in \mathcal{E}_{\mathcal{T}}$ of a set of given type $\mathcal{T} \in \{N, E, F, C\} \equiv$ {node, edge, face, cell} across the whole mesh provided that tag "State" exists. This algorithm corresponds to function *Mesh::AssignGlobalID* and creates the tag "GlobalID".

Algorithm 6.3 Global enumeration

$k = 0$
for all $e \in \mathcal{E}_\mathcal{T}$ **do**
  if State(e) is "owned" or "shared" **then**
    $k = k + 1$
  end if
end for
MPI_Exscan(k)
for all $e \in \mathcal{E}_\mathcal{T}$ **do**
  if State(e) is "owned" or "shared" **then**
    GlobalID(e) = k
    $k = k + 1$
  end if
end for

Packing and unpacking data. Let elements of set $\mathcal{E}_\mathcal{T}$ with type $\mathcal{T}$ be sorted by a unique identificator. We want the data related to tag "DataTag" to be packed to a byte array $\mathbb{B}$ for all elements. Algorithm 6.4 describes packing of the tag data to a byte array and is implemented in *Mesh::PackTagData*.

The algorithm exploits operations MPI_Pack($\cdot$, $\mathbb{B}$), MPI_Unpack($\cdot$, $\mathbb{B}$). Operation MPI_Pack($\cdot$, $\mathbb{B}$) places data to a byte array $\mathbb{B}$ following the steps: call MPI_Pack_size to retrieve size of packed data in bytes; if necessary, extend the byte array $\mathbb{B}$ to accommodate the data; call MPI_Pack to record the data into the byte array $\mathbb{B}$. Operation MPI_Unpack($\cdot$, $\mathbb{B}$) is defined via similar steps.

Algorithm 6.4 Pack tag data

create a byte array $\mathbb{C}$ and integer array $\mathbb{N}$.
reserve first two positions in $\mathbb{N}$
for all $\mathcal{T} \in \{N, E, F, C\}$ **do**
  for all $e \in \mathcal{E}_\mathcal{T}$ **do**
    if DataTag is sparse on $\mathcal{T}$ **then**
      if DataTag have data on e **then**
        add position of e in $\mathcal{E}_\mathcal{T}$ to $\mathbb{N}$
      else
        continue
      end if
    end if
    add DataTag(e) to $\mathbb{C}$
    if DataTag is of variable size **then**
      add sizeof(DataTag(e)) to $\mathbb{N}$
    end if
  end for
end for
$\mathbb{N}(0)$ = sizeof($\mathbb{N}$)
$\mathbb{N}(1)$ = sizeof($\mathbb{C}$)
MPI_Pack ($\mathbb{N}$, $\mathbb{B}$)
MPI_Pack ($\mathbb{C}$, $\mathbb{B}$)

Algorithm 6.5 describes unpacking of the tag data from a byte array and is implemented in *Mesh::UnpackTagData*.

Algorithm 6.5 Unpack tag data

```
unpack first two integers $S_1$ and $S_2$ from $\mathbb{B}$ by MPI_Unpack
resize $\mathbb{N}$ by $S_1 - 2$ and $\mathbb{C}$ by $S_2$
unpack following $S_1 - 2$ integers to $\mathbb{N}$ from $\mathbb{B}$ by MPI_Unpack ($\mathbb{N}$, $\mathbb{B}$)
unpack following $S_2$ bytes to $\mathbb{C}$ from $\mathbb{B}$ by MPI_Unpack ($\mathbb{C}$, $\mathbb{B}$)
for $\mathcal{T} \in \{N, E, F, C\}$ do
    if DataTag is sparse on $\mathcal{T}$ then        ▷ Data is present only on some elements of $\mathcal{E}_\mathcal{T}$
        for all $e \in \mathcal{E}_\mathcal{T}$ do
            delete data related to DataTag($e$)
        end for
        for all $n \in \mathbb{N}$ do
            $e = \mathcal{E}_\mathcal{T}(n)$
            if DataTag is of variable size then
                get next $n \in \mathbb{N}$
                resize DataTag($e$) by $n$
            end if
            put sizeof(DataTag($e$)) bytes from $\mathbb{C}$ to DataTag($e$)
        end for
    else        ▷ Data is present on all the elements of $\mathcal{E}_\mathcal{T}$
        for all $e \in \mathcal{E}_\mathcal{T}$ do
            if DataTag is of variable size then
                get next $n \in \mathbb{N}$
                resize DataTag($e$) by $n$
            end if
            put sizeof(DataTag($e$)) bytes from $\mathbb{C}$ to DataTag($e$)
        end for
    end if
end for
```

In implementation, both Algorithms 6.4 and 6.5 are more complicated since they should account for complex data types such as variables with variations and references to other elements. The last two algorithms require that the set of elements $\mathcal{E}$ is sorted according to an unique identificator, in order to arrange coincident sets on different processors. The unique identificator may be based on the global enumeration (Algorithm 6.3), on the geometric center of mass of every element, or on a hierarchy of mesh elements. Each identificator entails comparators of mesh elements *Mesh::GlobalIDComparator*, *Mesh::CentroidComparator*, *Mesh::HierarchyComparator*, respectively. Any of the comparators may be used to sort the set.

Packing and unpacking the elements. Now suppose we want to exchange between processors a set of elements $\mathcal{E}$ which is composed of subsets $\mathcal{E}_\mathcal{T}$ of elements of different types $\mathcal{T}$.

We define subtraction of one from a type $\mathcal{T}$, i.e., $\{C-1, F-1, E-1, N-1\} = \{F, E, N, \emptyset\}$, addition of one to a type: $\{C+1, F+1, E+1, N+1\} = \{\emptyset, C, F, E\}$, and denote by $\mathrm{adj}(\cdot, \mathcal{T})$ retrieval of all adjacent elements of type $\mathcal{T}$ from the set $\mathcal{E}$ or an element e. Let integer tag "PackID" store local enumeration of elements in $\mathcal{E}_{\mathcal{T}}$ and sort $(\cdot)$ represent sorting of elements by any available global identificator.

Algorithm 6.6 packs elements of $\mathcal{E}$ to a byte array $\mathbb{B}$ and places all information necessary to reconstruct the hierarchical mesh structure on a remote processor. It is followed by Algorithm 6.4 with all mesh data tags for elements in $\mathcal{E}$. The presented algorithm is a simplified version as it does not consider a hierarchy of sets of elements and sets themselves. This functionality of the algorithm is scattered across functions *Mesh::PackElementsGather*, *Mesh::PackElementsEnumerate*, *PackElementsData*.

Algorithm 6.6 Pack set of elements

for all $\mathcal{T} \in \{C, F, E\}$ **do** ▷ Extension of sets by dependency to lower adjacencies
    $\mathcal{E}_{\mathcal{T}-1} = \mathcal{E}_{\mathcal{T}-1} \cup \mathrm{adj}(\mathcal{E}_{\mathcal{T}}, \mathcal{T}-1)$
    for all $k \in [1, \mathrm{sizeof}(\mathcal{E}_{\mathcal{T}})]$ **do** ▷ Enumerate elements in $\mathcal{E}_{\mathcal{T}}$ and put result to tag PackID
        $\mathrm{PackID}(\mathcal{E}_{\mathcal{T}}(k)) = k$
    end for
end for
MPI_Pack (sizeof ($\mathcal{E}_N$) , $\mathbb{B}$)
for all $e \in \mathcal{E}_N$ **do** ▷ Pack all nodes
    if GlobalID exists **then**
        MPI_Pack (GlobalID(e), $\mathbb{B}$)
    end if
    MPI_Pack (Coords(e), $\mathbb{B}$)
end for
for all $\mathcal{T} \in \{E, F, C\}$ **do**
    MPI_Pack (sizeof ($\mathcal{E}_{\mathcal{T}}$) , $\mathbb{B}$)
    for all $e \in \mathcal{E}_{\mathcal{T}}$ **do**
        MPI_Pack (sizeof (adj ($e, \mathcal{T}-1$)) , $\mathbb{B}$)
        for all $q \in \mathrm{adj}(e, \mathcal{T}-1)$ **do**
            MPI_Pack (PackID(q), $\mathbb{B}$)
        end for
    end for
end for

Algorithm 6.7 unpacks elements from binary array and reconstructs a hierarchy of mesh elements. It avoids duplication of elements by performing searches among lists of adjacency elements for all types of elements except nodes. For nodes, we have to search among all already existing nodes in the mesh. To perform this efficiently, we use the binary search with available comparator for the mesh.

Algorithm 6.7 Unpack set of elements

```
form set N of all existing nodes in the mesh.
sort (N)                                         ▷ Sort according to available global identificator
MPI_Unpack(S_N, B)
for all k ∈ [1, S_N] do
    if GlobalID exists then
        MPI_Unpack(ID, B)
    end if
    MPI_Unpack(xyz, B)
    if GlobalID exists then                      ▷ N is sorted by GlobalID in this case
        binary search of node e in N by ID
    else                                         ▷ N is sorted by center of mass in this case
        binary search of node e in N by xyz
    end if
    if e = ∅ then
        create node e with coordinates xyz
    end if
    put e to set E_N
end for
for all T ∈ {E, F, C} do
    MPI_Unpack(S_T, B)
    for all k ∈ [1, S_T] do
        MPI_Unpack(S_A, B)
        create temporary set E_A of size S_A     ▷ E_A corresponds to adjacencies list
        for all m ∈ [1, S_A] do
            MPI_Unpack(p, B)
            get e = E_{T-1}(p)
            put e to E_adj
        end for
        if ∩_{e∈E_adj} adj (e, T) ≠ ∅ then       ▷ Element already exists
            e = ∩_{e∈E_adj} adj (e, T)
        else
            create element e by adjacent elements E_A
        end if
        put element e to set E_T
    end for
end for
```

Now, we proceed to main algorithms for maintenance of a distributed mesh.

Initialization of distributed mesh. Suppose we have a distributed mesh with geometric information only. Since all maintenance Algorithms need tags "Status", "Owner", and "Processors" to be defined on all elements, Algorithms 6.8 and 6.9 recover this information. Algorithm 6.8 finds an approximate set of processors $\mathcal{P}$ for a given processor p that should participate in interprocessor communication:

Algorithm 6.8 Approximate set of processors

compute local bounding box bb around all of the nodes
exchange bounding boxes BB between processors by MPI_Allgather
for all $b_k \in BB$ **do**
  if $b_k \cap bb \neq \emptyset$ **then**
    add k to $\mathcal{P}$
  end if
end for

Algorithm 6.9 computes "State" tag by "Processors" tag for element e on processor p_1:

Algorithm 6.9 Compute state

Owner(e) = min(Processors(e))
if Owner(e) = p_1 **then**
  if sizeof(Processors(e)) = 1 **then**
    State(e) = "owned"
  else
    State(e) = "shared"
  end if
else
  State(e) = "ghost"
end if

Algorithm 6.10 initializes tag "Processors" in nodes. The set $\mathcal{P}$ of interprocessor communication is updated by intersection of "Processors" data on nodes.

Algorithm 6.10 Initialization of nodes

sort($\mathcal{N}$) by coordinates
for all $e \in \mathcal{N}$ **do**
  for all $k \in [1, \text{sizeof}(BB)]$ **do**
    if Coords(e) $\in b_k \in BB$ **then**
      MPI_Pack(Coords(e), $\mathbb{B}_{k,out}$)
    end if
  end for
end for
exchange arrays $\mathbb{B}_{\mathcal{P},out}$ by Algorithm 6.2
for all $p \in \mathcal{P}$ **do**
  MPI_Unpack $(\mathcal{N}_p, \mathbb{B}_{p,in})$ ▷ Nodes from remote processor
  $\mathcal{N}_p = \mathcal{N}_p \cap \mathcal{N}$ ▷ Efficient for sorted arrays
  for all $e \in \mathcal{N}_p$ **do**
    add p to Processors(e)
  end for
end for

Algorithm 6.11 initializes "Processors" tag on edges, faces, and cells.

Algorithm 6.11 Initialization of edges, faces, and cells

for all $\mathcal{T} \in \{E, F, C\}$ **do**
  for all $e \in \mathcal{E}_{\mathcal{T}-1}$ **do**
    compute "State" of e by 6.9
  end for
  use Algorithm 6.3 to enumerate elements of type $\mathcal{E}_{\mathcal{T}-1}$
  for all $p \in \mathcal{P}$ **do**
    create temporary set $\mathcal{A}_p$ ▷ $\mathcal{A}_p$ corresponds to local estimate of shared elements
    for all $e \in \mathcal{E}_{\mathcal{T}}$ **do**
      for all $q \in \text{adj}\,(e, \mathcal{T} - 1)$ **do**
        if $p \in \text{Processors}(q)$ **then**
          add q to $\mathcal{A}_p$
        end if
      end for
    end for
    Pack $\mathcal{A}_p$ to $\mathbb{B}_{p,out}$ by Algorithm 6.6
  end for
  exchange arrays $\mathbb{B}_{\mathcal{P},out}$ by Algorithm 6.2
  for all $p \in \mathcal{P}$ **do**
    create temporary set $\mathcal{A}'_p$
    upack arrays $\mathbb{B}_{p,in}$ by Algorithm 6.7 with modification:
    if element exists, we don't create it
    if element exists, we add it to $\mathcal{A}'_p$
    for all $e \in \mathcal{A}_p \cap \mathcal{A}'_p$ **do**
      add p to $\text{Processors}(e)$
    end for
  end for
end for

Initialization of "Status" and "Owner" tags in all cells is provided by Algorithm 6.9. The ensemble of Algorithms 6.8, 6.9, 6.10, and 6.11 is implemented in *Mesh::ResolveShared*. Algorithm 6.9 does not necessarily rely on Algorithms 6.6 and 6.7, as it is not very efficient to reconstruct all the hierarchy for sets $\mathcal{A}_p$. It is sufficient to collect global identificators of existing adjacent elements and send them to a remote processor. In turn, remote processors perform binary search over their elements to find correspondence.

Synchronization. For any two processors p_1, p_2, the number of "ghost" elements on p_1 with "owner" equal to p_2 equals to the number of "shared" elements with "processors" containing p_1 on p_2 and vice versa. This allows us to gather "ghost" elements $\mathbb{G}_{p_2}$ on processor p_1 for processor p_2 and corresponding "shared" elements $\mathbb{S}_{p_1}$ on processor p_2 for processor p_1 and sort them in a common way. From the performance point of view, it will be convenient to gather and sort these arrays beforehand.

Algorithm 6.12 synchronizes the tag data with the above arrays. The algorithm is implemented via *Mesh::ExchangeData* (synchronous) or via a combination of *Mesh::ExchangeDataBegin* and *Mesh::ExchangeDataEnd* (asynchronous).

Algorithm 6.12 Synchronization

1: **for all** $p \in \mathcal{P}$ **do**
2: pack the data from $\mathbb{S}_p$ to $\mathbb{B}_{p,out}$ by Algorithm 6.6
3: **end for**
4: exchange arrays $\mathbb{B}_{\mathcal{P},out}$ by Algorithm 6.2
5: **for all** $p \in \mathcal{P}$ **do**
6: unpack the data from $\mathbb{B}_{p,in}$ to $\mathbb{G}_p$ by Algorithm 6.7
7: **end for**

Reduction algorithm runs Algorithm 6.12 in reverse. Reduction algorithm is needed, for example, to sum data from ghost elements to original element. This functionality is implemented via *Mesh::ReduceData* (synchronous) or via a combination of *Mesh::ReduceDataBegin* and *Mesh::ReduceDataEnd* (asynchronous). The function requires that the user provides a data accumulation operation as multiple copies of data will be received on shared elements. Examples of such accumulation operations are sum or maximum of received data.

Formation of layers of ghost elements. Collector algorithm creates ghost elements by user indication on local elements. The complication of the algorithm is the following. Consider three processors p_1, p_2, and p_3. Processor p_1 sends elements to processor p_2, and some of these elements belong to p_3. Processor p_3 cannot recognize that processor p_2 received its elements from processor p_1. This will result in broken synchronization since the number of shared and ghost elements will not match anymore. The remedy is to involve another round of communication in which processor p_2 informs processor p_3 about received elements.

Algorithm 6.13 Collector on processor p_1

for all $e \in \mathcal{E}$ **do**
    if e should be sent to p **then**
        add e to $\mathcal{E}_p$
        add p to $\mathcal{P}'$
    end if
end for
for all $p \in \mathcal{P}'$ **do**
    pack elements $\mathcal{E}_p$ to $\mathbb{B}_{p,out}$ by Algorithm 6.6
    pack the tag data of elements $\mathcal{E}_p$ to $\mathbb{B}_{p,out}$ by Algorithm 6.4
end for
exchange arrays $\mathbb{B}_{\mathcal{P}',out}$ by Algorithm 6.2
for all $p \in \mathcal{P}'$ **do**
    unpack elements $\mathcal{E}'_p$ to $\mathbb{B}_{p,in}$ by Algorithm 6.7
    unpack the tag data of elements $\mathcal{E}'_p$ to $\mathbb{B}_{p,in}$ by Algorithm 6.5
    for all $e \in \mathcal{E}'_p$ **do**
        if $p_1 \notin \text{Processors}(e)$ **then** ▷ p_1 is the current processor
            set $p_e = \text{Owner}(e)$
            add e to $\mathcal{E}''_{p_e}$
            add p_e to $\mathcal{P}''$
        end if
    end for
end for
for all $p \in \mathcal{P}''$ **do**
    pack elements $\mathcal{E}'_p$ to $\mathbb{B}_{p,out}$ by Algorithm 6.6
end for
exchange arrays $\mathbb{B}_{\mathcal{P}'',out}$ by Algorithm 6.2
for all $p \in \mathcal{P}''$ **do**
    unpack elements from $\mathbb{B}_{p,in}$ by Algorithm 6.7
    during unpacking add p to "Processors" tag of unpacked elements
end for
synchronize "Processors" tag by Algorithm 6.12

The user can indicate elements that should be ghosted by Algorithm 6.13. The algorithm is implemented in *Mesh::ExchangeMarked*. This function in INMOST can perform two kinds of actions, either ghosting or migration of elements. Migration is discussed later.

To create layers of ghost elements, we should define the set of faces that lay on the interface between local parts of the mesh of two processors p_1 and p_2. This set $\mathcal{S}_{p_2}$ is the starting set on p_1 from which the first layer of adjacent cells is considered. The set is also called the shared skin set.

If there are no ghost layers in the initial mesh distribution, then it is easy to determine the desired set. We gather into $\mathcal{S}_{p_2}$ all "shared" and "ghost" faces with processor p_2 on p_1. We will not consider more complicated case, for the sake of brevity. The algorithm that collects shared skin corresponds to function *Mesh::ComputeSharedSkinSet*.

Let $\overline{\text{adj}}(\mathcal{E}, \mathcal{T})$ represent the boundary set of adjacent elements of type $\mathcal{T}$ of set $\mathcal{E}$. It is obtained by selecting adjacent elements that are encountered only once, i.e.,

$$\overline{\mathrm{adj}}(\mathcal{E}, \mathcal{T}) = \left(\bigcup_{e \in \mathcal{E}} \mathrm{adj}(e, \mathcal{T}) \right) \setminus \left(\bigcup_{l \neq q \in \mathcal{E}} \mathrm{adj}(l, \mathcal{T}) \cap \mathrm{adj}(q, \mathcal{T}) \right),$$

and can be efficiently implemented using markers. Algorithm 6.14 is used to construct N ghost layers.

Algorithm 6.14 Ghosting

for all $p \in \mathcal{P}$ **do**
    for all $e \in \mathcal{S}_p$ **do**
        for all $q \in \mathrm{adj}(e, C)$ **do**
            if $p \notin \mathrm{Processors}(q)$ **then**
                mark q to be sent to p
                put q to set $\mathcal{L}_p^1$
            end if
        end for
    end for
end for
$\mathcal{S}_{\mathcal{P}}^1 = \mathcal{S}_{\mathcal{P}}$
for all $k \in [1, N]$ **do**
    use Algorithm 6.13 to exchange marked sets $\mathcal{L}_p^k$
    if $k < N$ **then** ▷ Assemble next layer
        for all $p \in \mathcal{P}$ **do**
            $\mathcal{S}_p^{k+1} = \overline{\mathrm{adj}}\left(\mathcal{L}_p^k, F\right)$ ▷ Boundary faces of current layer
            $\mathcal{S}_p^{k+1} = \mathcal{S}_p^{k+1} \setminus \mathcal{S}_p^k$ ▷ Exclude previous skin
            for all $q \in \mathrm{adj}\left(\mathcal{S}_p^{k+1}, C\right)$ **do**
                if $p \notin \mathrm{Processors}(q)$ **then**
                    mark q to be sent to p
                    put q to set $\mathcal{L}_p^{k+1}$
                end if
            end for
        end for
    end if
end for

Migration of elements is similar to Algorithm 6.13, but it does not require the second round of communication to inform owners. Instead, new "Processors" data is precomputed, given the new owner processor for each cell is provided. Algorithm 6.16 summarizes the steps required for computation of the "NewProcessors" data (including restoration of prescribed number of ghost layers). The algorithm misses handling of hierarchical sets and references to elements. The driver routine for migration is *Mesh::Redistribute*.

Summary of the main algorithms. Given the number of ghost layers and the mesh connectivity (neighbors through the faces, edges, or vertices), INMOST functions automatically compute and distribute ghost cells and organize exchanges of elements data from processors–owners to processors possessing the copies of the elements

Algorithm 6.15 Determine processors of an element e

$E_A = \text{adj}(e, \mathcal{T}(e) + 1)$

NewProcessors(e) $= \emptyset$

for all $q \in \mathcal{E}_A$ **do**

  NewProcessors(e) = NewProcessors(e) $\cup$ NewProcessors(q)

end for

Algorithm 6.16 Determining processors for migration of elements

compute partitioning for cells, i.e. define tag "NewOwner".

perform synchronization of "NewOwner" using Algorithm 6.12 for $\mathcal{T} = \{C\}$

for all $e \in \mathcal{E}_C$ **do**

  NewProcessors(e) = NewOwner(e)

end for

for all $\mathcal{T} \in \{F, E, N\}$ **do**

  for all $e \in \mathcal{E}_\mathcal{T}$ **do**

    calculate NewProcessors(e) using Algorithm 6.15

    set NewOwner(e) = min (NewProcessors(e))

  end for

end for

perform synchronization of "NewOwner" using Algorithm 6.12 for $\mathcal{T} = \{F, E, N\}$

if $N > 0$ **then** ▷ N layers are required after migration.

  perform reduction with union of "NewProcessors" data for $\mathcal{T} = \{F\}$

  perform synchronization of "NewProcessors" using Algorithm 6.12 for $\mathcal{T} = \{F\}$

  create shared skin sets $\mathcal{S}_\mathcal{P}$

  for all $e \in \mathcal{E}_F$ **do**

    if sizeof (NewProcessors(e)) $= 2$ **then**

      p_1 = NewProcessors(e)(1)

      p_2 = NewProcessors(e)(2)

      add e to $\mathcal{S}_{p_1}$ and $\mathcal{S}_{p_2}$

    end if

  end for

  for all $p \in \mathcal{P}$ **do**

    for all $e \in \mathcal{S}_p$ **do**

      for all $q \in \text{adj}(e, C)$ **do**

        if $p \notin$ NewProcessors(q) **then**

          put p to NewProcessors(q)

          put q to set $\mathcal{L}_p^1$

        end if

      end for

    end for

  end for

continued on the next page

[149]. Besides data update at ghost elements, reduction for data at ghost elements is possible as well: all processors that possess a ghost copy of an element send the data to the processor–owner of the element, where a user-defined function accumulates multiple data. Multiple ghost layers are handled thoroughly in INMOST, since cross-processor transfers occur commonly. An explicit ghost map may be defined for specific applications.

$\mathcal{S}_{\mathcal{P}}^1 = \mathcal{S}_{\mathcal{P}}$
for all $k \in [1, N]$ **do**
use Algorithm 6.13 to exchange marked sets $\mathcal{L}_p^k$
if $k < N$ **then** ▷ Assemble next layer
for all $p \in \mathcal{P}$ **do**
$\mathcal{S}_p^{k+1} = \overline{\text{adj}}\left(\mathcal{L}_p^k, F\right)$ ▷ Boundary faces of current layer
$\mathcal{S}_p^{k+1} = \mathcal{S}_p^{k+1} \setminus \mathcal{S}_p^k$ ▷ Exclude previous skin
for all $q \in \text{adj}\left(\mathcal{S}_p^{k+1}, C\right)$ **do**
if $p \notin \text{NewProcessors}(q)$ **then**
put p to NewProcessors(q)
put q to set $\mathcal{L}_p^{k+1}$
end if
end for
end for
end if
end for
perform reduction with union of "NewProcessors" data for $\mathcal{T} = \{C\}$
perform synchronization of "NewProcessors" using Algorithm 6.12 for $\mathcal{T} = \{C\}$
for all $\mathcal{T} \in \{F, E, N\}$ **do**
for all $e \in \mathcal{E}_{\mathcal{T}}$ **do**
calculate NewProcessors(e) using Algorithm 6.15
end for
end for
end if
perform reduction with union of "NewProcessors" data for $\mathcal{T} = \{N, E, F\}$
perform synchronization of "NewProcessors" using Algorithm 6.12 for $\mathcal{T} = \{N, E, F\}$
mark elements to be sent according to "NewProcessors"
use Algorithm 6.13 without second round
replace "Processors" with "NewProcessors"
replace "Owner" with "NewOwner"
recompute "State"

Deliberation of the main algorithms should include the following. Exchange of data "reference to an element" with a remote processor requires reconstruction of the referenced element on that processor. This implies ghosting of the referenced element and linking to this element on the remote processor. If the referenced element in its turn references some other elements, we may end up with the need to reconstruct the entire mesh on each processor, in order to maintain all the references. Referenced elements to be reconstructed on remote processors are serialized into binary buffers. Elements are unpacked from the buffer in the same order as they were serialized. To minimize the data exchange, we pack only global identificators and on the remote processor we use binary search of the element by global identificator among ghost or shared elements.

The method for exchange of "references to elements" is used in precomputing and maintaining a stencil of a discretization scheme and during mesh balancing. Further, we deliberate on the complexity of handling hierarchy of trees in parallel. We also

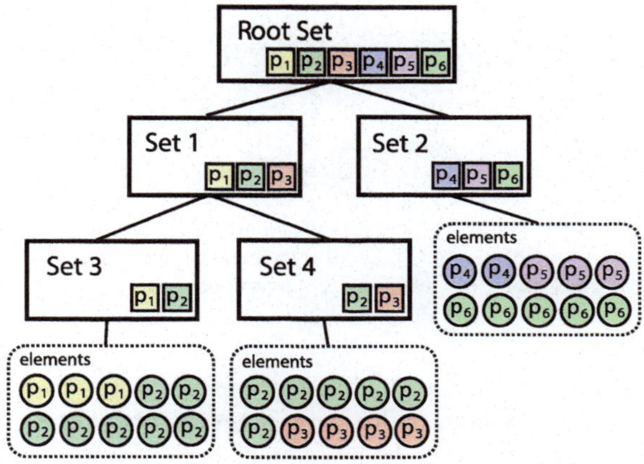

Fig. 6.14 Distribution of set hierarchy in parallel

detail balancing for both, the mesh overall, and ghost layers in particular. All of this is required in context of aggressive parallel dynamic remeshing.

6.3.1 Parallel Local Mesh Modifications

Parallel hierarchical sets. Since mesh elements may be organized into sets of elements possessing a tree structure (Fig. 6.2), it is important to handle in parallel the hierarchy of sets and their elements. Separation of elements between processors that belong to a single set should not require duplication of all the elements on each of the processors (Fig. 6.15). The following rules provide transparent logic on how sets and their elements are shared and exchanged in parallel (Fig. 6.14):

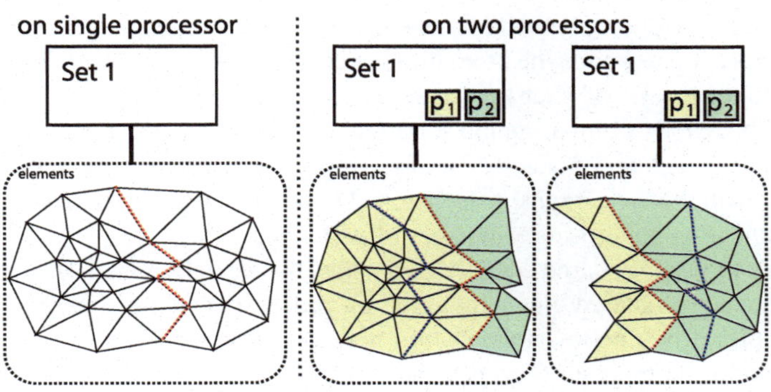

Fig. 6.15 Handling the elements of the set on single processor (left) and two processors (right) for the mesh with one layer of ghost elements. The same "Set 1" shared between both processors contains only elements present on each of the processors

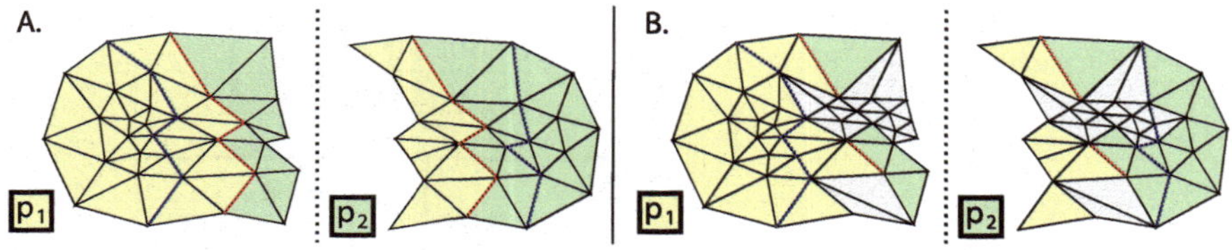

Fig. 6.16 Initial grid, the yellow and green color correspond to elements owned by processors p_1 and p_2, respectively (**a**). Local refinement of the grid and edge swap results in loss of ownership information for new elements, colored in gray (**b**)

- the set is uniquely defined by a name;
- the set belongs to the union of processors of its elements;
- a processor with the lowest rank is declared to be the owner of the set;
- when a hierarchical set is sent to a remote processor, its parent set has to be sent as well, so that entire upward hierarchy is reconstructed on that remote processor;
- a hierarchical set on a processor has to know all the links to elements and child sets that are present on the processor.

Information on set hierarchy and connections to elements is serialized into a binary buffer and, after sending it to a remote processor, is reconstructed from the buffer on that remote processor.

Modifications in parallel. Parallel local modifications (e.g., refinement or coarsening) of a general polyhedral mesh distributed among several processors require accurate handling. We assume that modifications are synchronized between processors, i.e., the modification on different processors should not produce topologically inconsistent and geometrically non-matching meshes. Once a distributed mesh is modified, the parallel consistency of the modified mesh has to be established.

Parallel state resolution after modification. The distributed modified mesh is obtained by deleting a part of the distributed input mesh (Fig. 6.16a) and adding new elements (Fig. 6.16b). According to Algorithm 6.10, in order to resolve geometrical correspondence of new elements on one processor to elements on the other processors, each processor builds the bounding boxes around all the new nodes. It collects, sorts by coordinates, and exchanges all its new nodes that fall into other processor's bounding boxes. By comparing the sorted lists of nodal coordinates, each processor determines for each new node the list of processors that have a copy of the node. The processor with the lowest rank is determined to be the owner of the new node. Thus, a communication pattern for exchanges of nodes is set.

In order to resolve shared edges, faces, and cells, we determine in Algorithm 6.11 a candidate set of shared elements based on processors' list of lower adjacencies. For example, if two nodes, node n_1 and node n_2 of an edge, belong to processors $\{p_1, p_2\}$ on local processor (say, p_1), then we assume that an edge $e_{12} = (n_1, n_2)$ on p_1 may also belong to both of these processors $\{p_1, p_2\}$. This is not necessarily true, as the edge e_{12} may not present on p_2. Then, the processors exchange the candidate sets and check elements which really exist on remote processors. As a result, we get

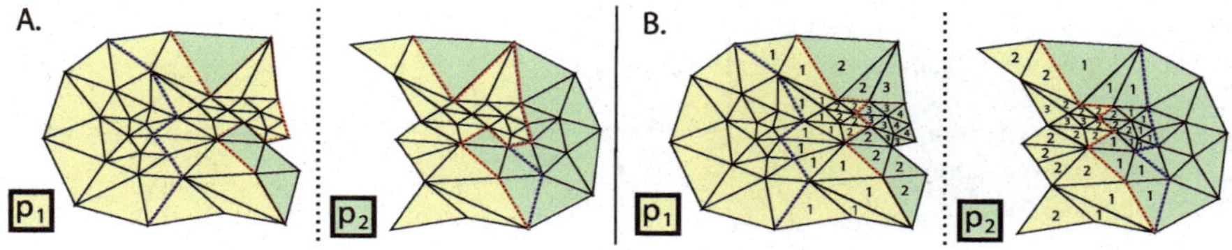

Fig. 6.17 Resolution of shared elements establishes ownership information, required for data exchange, but attributes all shared elements to processor p_1 (**a**). Assignment of element's owner processor by the minimal distance in the graph of shared and ghost elements connected by nodes (**b**)

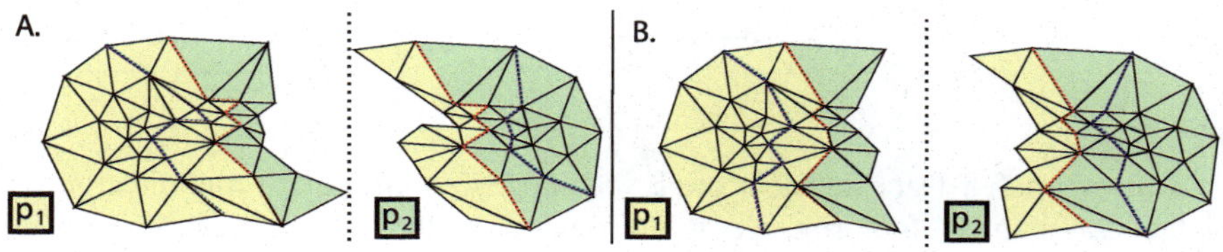

Fig. 6.18 Reconstruction of ghost layers leads to removal of unnecessary elements and addition of one element (**a**), and the resulting mesh has 43 and 28 elements owned by processor p_1 and p_2, respectively. Load balancing results in 35 and 36 elements owned by processor p_1 and p_2, respectively (**b**)

processors' list for edges, then faces and, finally, cells. The procedure is illustrated in Fig. 6.17a.

Balancing ghost layers. The ghost and shared elements are then equilibrated between the processors by computing the graph distance to elements owned by each processor. The graph nodes are formed by shared and ghost elements, and the graph edges are defined by lower adjacency elements prescribed by the user. To compute the minimal distance, we perform the data reduction operation from ghost elements to shared elements and then synchronize the reduced data among the processors. A processor with the minimal distance to the ghost or shared element in the graph is assigned to be the owner of that element. In case of equal distances for several processors, the owner is defined to be a processor with the lowest rank. The result of the procedure is illustrated in Fig. 6.17b.

Restoration of ghost layers. It remains to add (or remove) ghost elements in places where there is insufficient (or excessive) number of ghost layers. The procedure is equivalent to ghosting Algorithm 6.14. To this end, we compute a set of faces shared by two cells owned by different processors. The 2D analog is shown in Fig. 6.18a where such set of edges for all cells is marked by the red dotted line. Starting from the set of faces, we collect a layer of owned and shared cells and, if necessary, exchange them with the processor sharing these faces. The procedure can be repeated for the outer adjacent elements of the collected layer until the required number of ghost layers is reached. Unvisited elements are to be deleted. The result is illustrated in Fig. 6.18a.

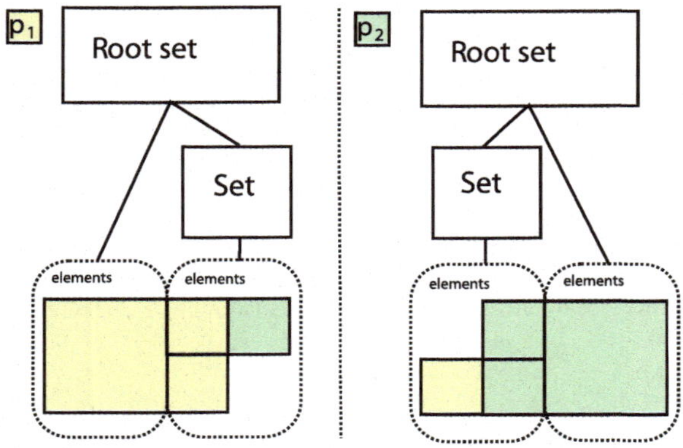

Fig. 6.19 A single cell, required for coarsening step, is missing on each of the processors. The situation requires synchronization of the elements of the sets

Data transfer. Data transfers often accompany parallel mesh modification. To reduce the memory footprint, we do not require to store the original mesh before modification. They are implemented in INMOST via mesh modification epochs, see Algorithm 6.1.

Summary. INMOST-based parallel implementation of local refinement and coarsening of general meshes maintains the number of ghost layers requested by the user and handles the cases of missing cells that are to be coarsened (Fig. 6.19). To ensure topological consistency, the INMOST tools synchronize between processors' mesh edge splitting, references to hanging nodes, and parent sets of cells.

Local mesh modification may change a large portion of the mesh even after one sweep of cells' modification. Therefore, local mesh refinement and coarsening may result in significant disbalance in the number of mesh cells (the degrees of freedom in the cell-centered FV discretizations) at processors, rf. Fig. 6.18a. Balancing the mesh involves computation of a new mesh distribution and migration of mesh elements, Fig. 6.18b.

6.3.2 *Mesh Balancing and Redistribution*

INMOST provides flexible interface for mesh repartitioning and redistribution. The user may choose one of internal partitioners, external partitioners Zoltan [37] and Parmetis [93], or provide the partitioning map explicitly. The redistribution process effectively utilizes the ghost layers if they were computed in advance and ensures that the ghost layers' structure is recovered after the redistribution.

In case of local modifications of a mesh distributed among given processors, new distribution of mesh cells should correlate with the former distribution, in order to minimize the communication load and computational work for mesh reconstruction. We employ the K-means clustering algorithm [79]. On the initialization step, the initial mesh distribution defines clusters of mesh cells owned by each processor.

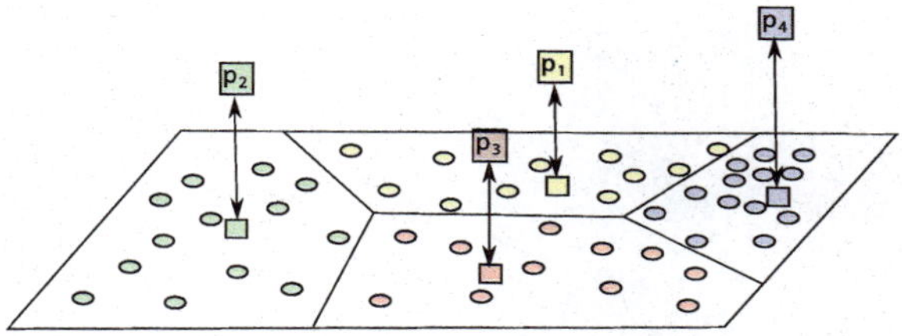

Fig. 6.20 The K-means clustering algorithm for a set of cells with the additional coordinate for cluster centers. The additional coordinate is proportional to the number of cells assigned to each cluster and the diagonal of the bounding box of the cluster submesh

Centers of clusters move smoothly as the mesh is locally modified. Yet, the numbers of cells assigned to each cluster mismatch. To alleviate the problem, we introduce the fourth (third in 2D) coordinate to the centers of clusters as illustrated in Fig. 6.20: the additional coordinate is proportional to the number of cells assigned to the cluster and to the diagonal of the bounding box of the cluster submesh.

Application of the K-means clustering algorithm to the mesh with the additional coordinate rebalances the mesh among processors better.

Once the new mesh distribution is computed, migration of elements is performed by serializing information into binary buffer as in Algorithm 6.4. For elements, we serialize their lower adjacency list and all their data. For sets of elements, we serialize their name, data, upward graph tree structure, and connections to child sets and set elements, which are sent to or present on the remote processor. After serialization, all the data related to migrated elements is deleted on local processor.

We summarize the INMOST operations handling distributed mesh data:

- distribute a mesh between processors;
- specify ghost elements;
- exchange tag data for ghost elements;
- reduce tag data for ghost elements;
- refine locally a mesh;
- coarsen locally a mesh;
- redistribute a mesh between processors;
- transfer data between meshes.

6.3.3 Numerical Example

As the initial distributed mesh, we consider a single-layer hexagonal prismatic mesh in the unit cube $\Omega = [0, 1]^3$. The mesh is refined by the two-level refinement in a moving ring $0.125 < \sqrt{(x_c - x)^2 + (y_c - y)^2} < 0.175$ with parameters:

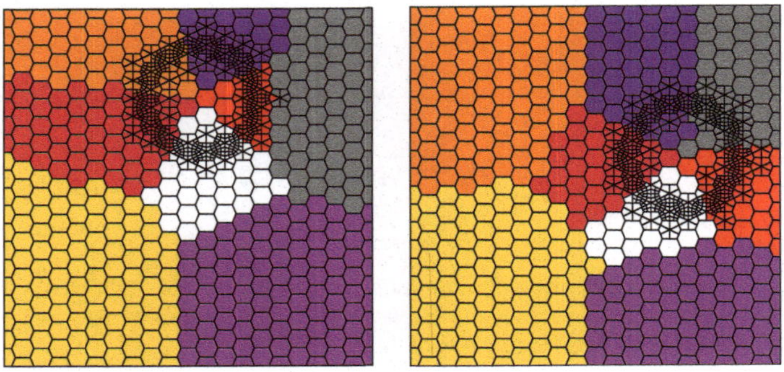

Fig. 6.21 Parallel mesh adaptation of $20 \times 20 \times 1$ hexagonal mesh, distributed among eight processors, at steps $k = 1$ (left) and $k = 8$ (right)

$$
\begin{aligned}
x_c &= \frac{1}{2} + \frac{1}{4}\cos\left(\frac{k\pi}{20}\right),\\
y_c &= \frac{1}{2} + \frac{1}{4}\sin\left(\frac{k\pi}{20}\right),
\end{aligned}
\tag{6.1}
$$

where k is the time step number. The meshes due to the adaptive refinement of the original $20 \times 20 \times 1$ hexagonal prismatic grid at $k = 1$ and $k = 8$ are shown in Fig. 6.21. During refinement, some cells are split in z-direction as well, see Fig. 6.12.

For examination of the parallel performance, we start with the finer hexagonal $120 \times 120 \times 1$ grid containing 14580 prismatic cells. The region of interest (the ring) makes two revolutions around the central z-axis of the domain within 80 time steps. During each step, more than a half ($\approx$77%) of mesh elements are changed. Each two-level refinement increased the number of cells to $\approx$115000 and each two-level coarsening decrease it to $\approx$83000. Approximately 64000 cells are modified during each step.

Table 6.1 collects time measurements of all 80 steps, obtained on INM RAS cluster [6]. The refinement and coarsening work is not distributed equally among processors, and the method demonstrates reduced scalability.

Table 6.1 Performance of mesh adaptation test on different numbers of processors

Processors	1	12	24	48	96
Refinement, s	699.409	319.795	243.311	152.678	120.228
Coarsening, s	1034.540	325.289	242.216	144.145	111.534
Balancing, s	0.000	47.404	40.089	29.773	28.001
Total, s	1734.661	692.512	525.647	326.646	259.852

6.4 Linear System Assembly

All discretization methods considered in this book (as well as finite difference and finite element methods) result in a sparse matrix $\mathbb{A}$ of size $n \times n$ with a mean number m of non-zero elements per row which requires to keep $O(m \times n)$ values in the computer memory. The portrait of a sparse matrix $\mathbb{A}$ obtained from FV discretization of the diffusion problem on unstructured mesh is demonstrated in Fig. 6.22.

A usual approach to store a sparse matrix $\mathbb{A}$ is the compressed storage row (CSR) format. CSR format defines three contiguous arrays a, ia, and ja, where ia sets intervals of indices for each row, ja are column indices in each row, and a are the corresponding values. The drawback of such storage scheme is that it is not easy to insert a new non-zero value $a_{i,j}$ into the ith row and the jth column of matrix $\mathbb{A}$. Such operation requires to enlarge a and ja arrays (in the worst case at cost of $O(mn)$ operations), search for the insertion position (binary search over intervals in ia and column indices in ja takes $O(\log_2(n)\log_2(m))$ operations), and shift all entries that should stand after the newly inserted element (in the worst case at cost of $O(mn)$ operations). Such insertion operation is prohibitively costly in CSR format as it may require to move almost all elements of matrix $\mathbb{A}$. However, the CSR format is useful for methods which form the matrix row by row (e.g., row-wise incomplete LU factorization) or add matrix entries into a preliminary computed sparsity structure. In the last case, the solution to the problem of insertion a new non-zero entry is to preallocate enough storage for arrays, setup indices for intervals ia according to the starting point for each row, and introduce another array ia_e pointing to the last inserted index

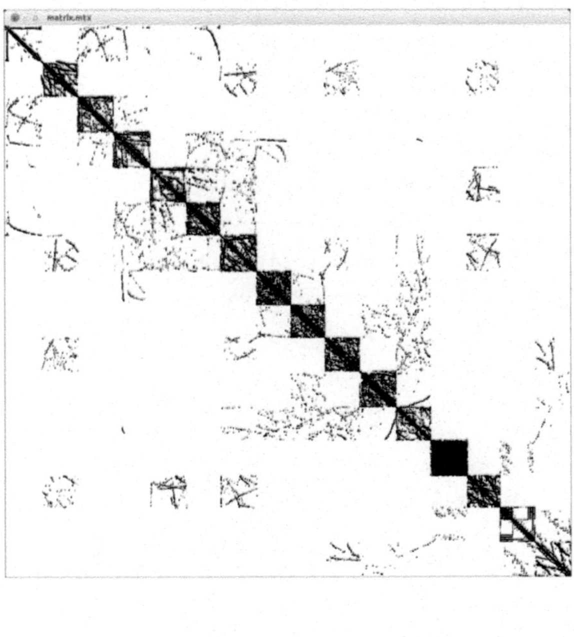

Fig. 6.22 Spy of the sparse linear system distributed among 16 processors for the diffusion problem. The picture is obtained using built-in example *DrawMatrix*

in each row. Then for the element insertion we have to search for the ith row position and then for the proper jth column position, shift elements that should appear after the jth column in the current row, and increment ia_e entry corresponding to the row. This reduces the insertion cost down to $O(\log_2(m)\log_2(n) + m)$ operations since $O(m)$ operations are required to shift elements in the row. However, the knowledge of non-zero pattern may be not available during incremental assembly of matrix $\mathbb{A}$, since each part of a multi-physics model that contributes to $\mathbb{A}$ may be implemented independently as a black box.

A sparse matrix in INMOST is implemented in class *Sparse::Matrix* that internally contains an array of objects of class *Sparse::Row*. On each processor, the rows are indexed according to the interval of unknowns of local processor partition. Each object of class *Sparse::Row* contains a dynamically re-sizable array of pairs of column index and corresponding coefficient. The column indices are not required to be inserted in the sorted order and require a linear search to avoid column index duplication on insertion. This results in $O(m)$ cost for $a_{i,j}$ insertion. The structure heavily uses memory allocation and reallocation that reduces the overall performance; however, minimal knowledge on the pattern of the matrix is required. As a result, the matrix can be easily extended either by introduction of new elements into existing rows or by attaching new rows. Class *Sparse::Matrix* additionally realizes method *RowVec* of multiplication of a sparse matrix and a dense vector. Such operation is required in parallel iterative solvers.

The functionality to store the Hessian of a vector function (3D array of second derivatives) is also available in INMOST and is realized through classes *INMOST::HessianMatrix* and *INMOST::HessianRow*. A similar structure is used for the Hessian storage: entries in each row contain two indices and single coefficient. The class additionally realizes multiplication of a Hessian and a vector. Such operation allows the user to implement the Halley method (6.15) for the iterative solution of a nonlinear system.

6.5 INMOST Linear Solvers

The basic internal linear solvers incorporated into INMOST platform are based on the combination of the following components:

- *Iterative method.* Preconditioned Krylov solver for non-symmetric matrices;
- *Parallelization.* Domain decomposition;
- *Preconditioning.* Incomplete LU factorization;
- *Preprocessing.* Matrix rescaling, non-symmetric maximum transversal permutation, symmetric fill-in reduction permutation;
- *Pivoting.* Delayed multi-level factorization.

Below, we address these components in detail.

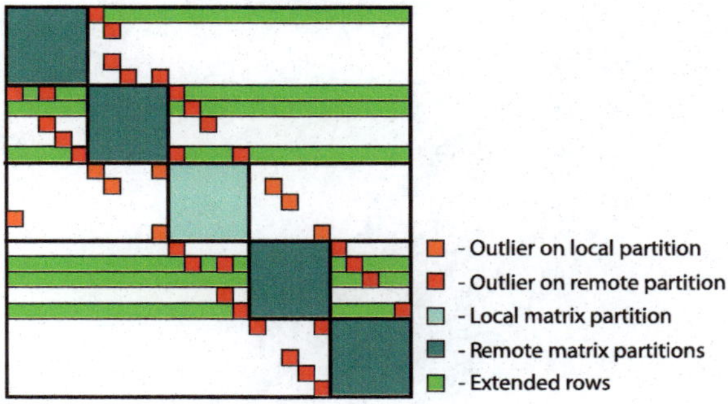

Fig. 6.23 Sparse matrix extension procedure

6.5.1 Parallel Iterative Method

The basic iterative method of INMOST is the preconditioned biconjugate gradient stabilized method BiCGStab(ℓ) [135]. This method optionally performs ℓ BiCG steps and fits a polynomial function to accelerate convergence of preconditioned residual. The BiCGStab(ℓ) parallelization is straightforward as it only requires to accumulate sums of scalar products computed on each processor and synchronize vector elements after matrix–vector multiplication.

Parallel implementation of the preconditioner is based on the combination of an incomplete LU factorization and the restricted additive Schwarz method with a user-specified overlapping parameter. Overlapping of a local matrix and a local vector requires their extension by data from adjacent processors. To construct the overlap, the sparsity pattern is analyzed for the global column indices that lay outside of the local processor and the local matrix is augmented with the rows from remote processors as illustrated in Fig. 6.23. The procedure of matrix extension is repeated q times where q is the user given overlap parameter. Zero overlap corresponds to the block Jacobi method.

Upon local matrix extension, each processor performs its incomplete LU factorization. All the column indices that fall outside of extended matrix partition are ignored. In BiCGStab(ℓ) iterations, after solution step with the preconditioner, the restriction to the extended preconditioned vector is applied.

All parallel operations for the restricted Additive Schwarz method are implemented in class *Solver::OrderInfo*.

6.5.2 Preprocessing

Prior to the factorization, the matrix is preprocessed by reordering and rescaling in three steps that are summarized as follows:

- Non-symmetric permutation for the static choice of pivoting sequence;
- Symmetric permutation for fill-in reduction;
- Rescaling to improve the condition number and dropping strategy.

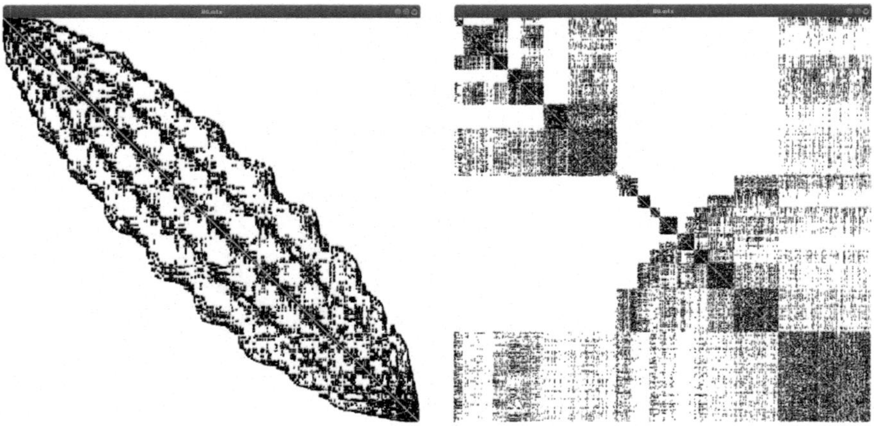

Fig. 6.24 Fill-in reduction reordering using built-in reverse Cuthill–Mckee method (left) and external node dissection method from METIS library (right). The matrix corresponds to the last Newton iteration of the FV Navier–Stokes solver for a Pousielle flow in a coarse pipe model. The non-zeros of the matrix are magnified for visibility

For static pivoting, we find the maximum transversal permutation [119]. The result is the reordering that maximizes the product of elements on diagonal and the rescaling that leads to I-dominant matrix. For efficient implementation of static pivoting for sparse matrices, we use the Dijkstra algorithm [60] with binary heap in dense array and dense linked lists. Such permutation is known to reduce the necessity for pivoting. However, small diagonal values or badly conditioned factors may still be encountered during factorization. To this end, we use dynamic pivoting detailed in the next subsection.

Symmetric permutations are used to reduce the number of non-zeros in the triangular factors. Two variants of such permutations are available in INMOST. The first variant is based on reverse Cuthill–Mckee reordering [50]. The second variant relies on the node dissection algorithm from METIS library [92]. Usually, the computation of reverse Cuthill–Mckee reordering is much cheaper although the fill-in reduction is much better with the node dissection algorithm from METIS. We recommend the first variant to users sensitive to license restrictions of the second variant. The result of application of the methods to the matrix produced by the last Newton iteration of the FV Navier–Stokes solver for a Pousielle flow in a coarse pipe model is illustrated in Fig. 6.24.

Once the permutations are over, we rescale the matrix. Two rescaling strategies provided by INMOST improve the condition number of the matrix to be factorized. The first strategy is the iterative equilibration of row and column norms to unity [87, 136]. The second strategy is the transformation of the matrix into I-dominant matrix which has unit values on its diagonal and off-diagonal entries with modulii not exceeding 1. The I-dominant matrix rescaling is the by-product of the algorithm that finds the maximal transversal and can be further improved iteratively.

In summary, the incomplete LU factorization is performed for preprocessed matrix $\tilde{\mathbb{A}}$ rather than for the original matrix $\mathbb{A}$:

$$\tilde{\mathbb{A}} = \mathbb{D}_L \mathbb{P}_R \mathbb{A} \mathbb{P}_C \mathbb{D}_R = \tilde{\mathbb{L}}\tilde{\mathbb{D}}\tilde{\mathbb{U}} + \mathcal{E},$$

where $\mathbb{D}_L$ and $\mathbb{D}_R$ are left and right diagonal rescaling matrices and $\mathbb{P}_R$ and $\mathbb{P}_C$ are row and column permutation matrices and $\mathcal{E}$ is the factorization error, $\tilde{\mathbb{L}}$ and $\tilde{\mathbb{U}}$ correspond to the triangular factors $\mathbb{L}$ and $\mathbb{U}$ which form the preconditioner. After the factorization is completed, the incomplete factors are rescaled:

$$\mathbb{L} = \mathbb{D}_L^{-1}\tilde{\mathbb{L}}\mathbb{D}_L, \quad \mathbb{D} = \mathbb{D}_L^{-1}\tilde{\mathbb{D}}\mathbb{D}_R^{-1}, \quad \mathbb{U} = \mathbb{D}_R\tilde{\mathbb{U}}\mathbb{D}_R^{-1}.$$

Applying the preconditioner to a vector b requires the solution of the system:

$$\mathbb{P}_R^T \mathbb{L}\mathbb{D}\mathbb{U}\mathbb{P}_C^T \mathbf{x} = \mathbf{b},$$

which is a two-stage process, (1) reordering by $\mathbb{P}_\mathbb{R}$ and the forward substitution

$$\mathbb{P}_R^T \mathbb{L}\mathbb{D}\mathbf{y} = \mathbf{b} \implies \mathbf{y} = \mathbb{D}^{-1}\mathbb{L}^{-1}\mathbb{P}_R \mathbf{b},$$

and (2) the backward substitution and reordering by $\mathbb{P}_C$

$$\mathbb{U}\mathbb{P}_C^T \mathbf{x} = \mathbf{y} \implies \mathbf{x} = \mathbb{P}_C \mathbb{U}^{-1}\mathbf{y}.$$

The number of iterations and the density of $\mathbb{L}$ and $\mathbb{U}$ factors may suggest to perform reordering of elements in $\mathbb{L}$ and $\mathbb{U}$ factors.

To improve the robustness of the incomplete factorization in the future, we plan to add local 2×2 pivoting [43] and block-structured elimination. To accelerate the process on computers with shared memory, we plan to add reordering for OpenMP parallelization [20].

6.5.3 Preconditioning

INMOST provides two variants of incomplete LU factorization. These are row-wise incomplete LU method and Crout incomplete LU method [104]. The difference in the methods is in the access pattern for matrix $\mathbb{A}$ to be factorized and construction of $\mathbb{L}$ and $\mathbb{U}$ triangular factors as illustrated in Fig. 6.25.

Both factorization methods use the second-order dual-threshold dropping strategy with τ_1 and τ_2 parameters following [88]. Usually, factorization is applied to preliminary preprocessed matrices, for which $\tau_2 = O(\tau_1^2)$ or $1 \gg \tau_1 \gg \tau_2 > 0$. In comparison with the conventional threshold LU factorization ILU(τ), the second-order accurate ILU2(τ_1, τ_2) factorization methods provide better preconditioner quality and better convergence of Krylov subspace iterations. For stiff SPD matrices and

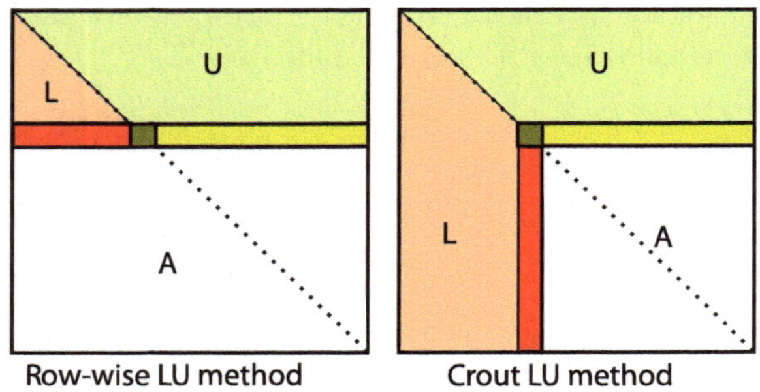

Fig. 6.25 Factorization pattern for row-wise LU and Crout LU methods

$\tau_2 = \tau_1^2$, the quality of the preconditioner is comparable to ILU(τ_2) and the preconditioner density is comparable to ILU(τ_1).

The essence of the second-order factorization method consists in storage of two versions of factored matrices $\mathbb{L}, \hat{\mathbb{L}}$ and $\mathbb{U}, \hat{\mathbb{U}}$. Matrices $\mathbb{L}$ and $\mathbb{U}$ contain elements whose modulii exceed the first threshold τ_1 and $\hat{\mathbb{L}}$ and $\hat{\mathbb{U}}$ contain elements whose modulii exceed the second threshold τ_2. During factorization, elimination is performed with products of $\mathbb{L}$ with $\mathbb{U}$ and $\hat{\mathbb{U}}$, products of $\mathbb{U}$ with $\mathbb{L}$ and $\hat{\mathbb{L}}$, i.e., contributions of the product of $\hat{\mathbb{L}}$ and $\hat{\mathbb{U}}$ are ignored. This allows us to improve significantly the quality of $\mathbb{L}$ and $\mathbb{U}$ factors approaching them closer to the triangular factors of the exact factorization. Once the factorization is completed, $\hat{\mathbb{L}}$ and $\hat{\mathbb{U}}$ factors are abandoned and the iterations proceed with the preconditioner involving factors $\mathbb{L}^{-1}$ and $\mathbb{U}^{-1}$.

On input, both incomplete LU factorization methods require

- sparse unsymmetric matrix $\mathbb{A}$ with non-negative diagonal elements;
- left and right scaling matrices $\mathbb{D}_L$ and $\mathbb{D}_R$;
- truncation threshold parameters $1 \gg \tau_1 \gg \tau_2 > 0$.

The row-wise incomplete LU method factorizes the matrix row by row and incrementally constructs row of $\mathbb{U}$ and row of $\mathbb{L}$. The row-wise method is computationally efficient for matrix $\mathbb{A}$ stored by rows. Note that the row-wise variant of the two-parameter ILU factorization drops off elements of matrix $\hat{\mathbb{L}}$ after processing the ith row of $\mathbb{A}$. Algorithm 6.17 corresponds to the row-wise incomplete LU method.

In Algorithm 6.17, matrix $\mathbb{A}$ and factors $\mathbb{L}$ and $\mathbb{U}$ are stored by rows. All the inner loops are made along the sparsity structure indices. Other loops over row accumulator vector v are based on linked list data structure. For details, we refer to [97].

In the Crout LU method, one has to access simultaneously both rows and columns of matrix $\mathbb{A}$ as well as rows and columns of $\mathbb{L}$ and $\mathbb{U}$ factors. Matrix $\mathbb{L}$ is constructed and stored by columns, whereas matrix $\mathbb{U}$ is constructed and stored by rows. Therefore, additional data structures are needed for traversal of $\mathbb{L}$ by rows and for traversal of $\mathbb{U}$ by columns. The elimination pattern is illustrated in Fig. 6.26. In the Crout LU method, we store the diagonal part separately, i.e., we compute LDU factorization, where triangular factors $\mathbb{L}$ and $\mathbb{U}$ have unit values on their diagonals.

The elimination process corresponds to the addition of multiple sparse vectors with a coefficient. Both factorization methods utilize an ordered dense linked list

Algorithm 6.17 ILU2: Second-order row-wise incomplete LU factorization

for $i \in [1:n]$ **do** ▷ Main loop by rows of $\mathbb{A}$
  $v = (\mathbb{D}_L \mathbb{A} \mathbb{D}_R)_{i,[1:n]}$
  for $k \in [1:i-1] : v_k \neq 0$ **do** ▷ Perform elimination
    $v_k = v_k / \mathbb{U}_{kk}$
    if $|v_k| > \tau_2$ **then** $v_{[k+1:n]} = v_{[k+1:n]} - v_k \mathbb{U}_{k,[k+1:n]}$ **end if**
    if $|v_k| > \tau_1$ **then** $v_{[k+1:n]} = v_{[k+1:n]} - v_k \hat{\mathbb{U}}_{k,[k+1:n]}$ **end if**
  end for
  $\mathbb{L}_{ii} = \max(\tau_2, \max_{k \in [i:n]}(|v_k|))$
  $v_{[i:n]} = v_{[i:n]} / \mathbb{L}_{ii}$
  for $j \in [1:i-1] : v_j \neq 0$ **do** ▷ Compute the ith $\mathbb{L}$ row
    if $|v_j| > \tau_1$ **then** $\mathbb{L}_{ij} = v_j$ **end if**
  end for
  $v_i = \text{sign}(v_i) \max(\tau_2, |v_i|)$
  $\mathbb{U}_{ii} = v_i$
  for $j \in [i+1:n] : v_j \neq 0$ **do** ▷ Compute the ith $\mathbb{U}$ and $\hat{\mathbb{U}}$ rows
    if $|v_j| > \tau_1$ **then** $\mathbb{U}_{ij} = v_j$
    else if $|v_j| > \tau_2$ **then** $\hat{\mathbb{U}}_{ij} = v_j$
    end if
  end for
end for
for $i \in [1, n]$ **do** ▷ Final $\mathbb{L}$ and $\mathbb{U}$ rescaling
  for $j \in [1:i] : \mathbb{L}_{ij} \neq 0$ **do**
    $\mathbb{L}_{ij} = \mathbb{L}_{ij} / \mathbb{D}_{L,ii}$
  end for
  for $j \in [i:n] : \mathbb{U}_{ij} \neq 0$ **do**
    $\mathbb{U}_{ij} = \mathbb{U}_{ij} / \mathbb{D}_{R,jj}$
  end for
end for

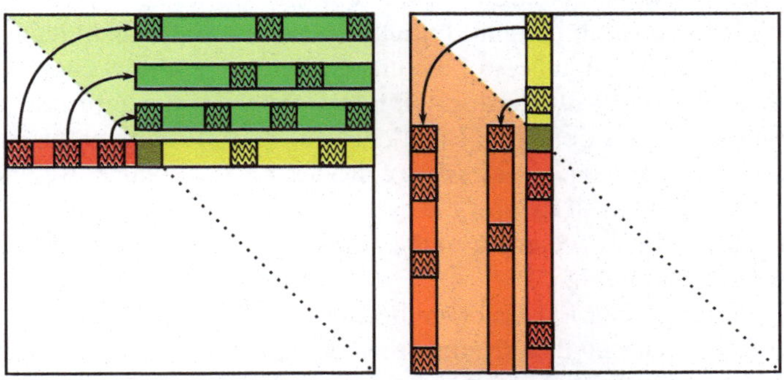

Fig. 6.26 Elimination patterns for row of factor $\mathbb{U}$ and column of factor $\mathbb{L}$ in the Crout LU algorithm

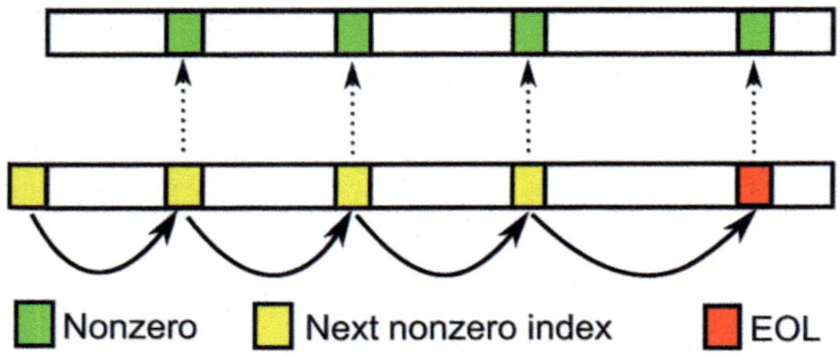

Fig. 6.27 Illustration for ordered dense linked list data structure

data structure to perform the elimination. The structure requires two arrays of indices and values of the size that covers all unknowns. Array of indices contains the next non-zero position with a larger index than the current position index. The array of values contains non-zero values. For insertion of a new element, the search for the proper position is required. However, insertion of a sparse vector with ordered column indices is performed efficiently, since unwinding to the beginning of the linked list is not needed for insertion of a next element from the sparse vector. The data structure is illustrated in Fig. 6.27. It is also possible to use unordered linked list with $O(1)$ insertion cost. Such structure is further explained and used in Sect. 6.6. The ordered dense linked list results in ordered column indices in sparse vectors which is required in the Crout ILU method. In parallel implementation, the sizes of arrays for a dense linked list are equal to the sizes of local overlapped matrices.

The data structure for the Crout LU factorization uses three additional arrays of indices. The first array I_1 points to the current non-zero in the row, the second array I_2 realizes linked list of row positions for which I_1 points to the element with the same column index, and the third array I_3 contains positions of the first elements in the linked list for each column. Initially, I_1 points to the first element in each row. Considering column indices of the first elements in each row, we construct the ordered linked list I_2 and add starting points to I_3. The first non-zero entry of the first column is located in position $I_1[I_2[I_3[1]]]$, the next non-zero (if present) is located at $I_1[I_2[I_2[I_3[1]]]]$, and so on. If we have to consider the next column, we first advance positions in I_1 array for all rows of the current column and update the linked list I_2 and starting positions I_3 according to the new column positions in these rows. The process is illustrated in Fig. 6.28.

The incomplete Crout LU factorization suggests additional improvement of the dropping strategy. During factorization, the condition number of the inverse factors $|\mathbb{L}^{-1}|$ and $|\mathbb{U}^{-1}|$ is estimated and thresholds τ_1 and τ_2 are adapted according to the estimates [35]. Such dropping strategy increases the density of the preconditioner for the same parameters τ_1 and τ_2. However, the method becomes more robust with much larger values τ_1 and τ_2. The cost for the condition number estimation is the solution procedure with two right-hand side vectors per factor to be performed along the factorization.

Algorithm 6.18 ILUC2: Second-order Crout incomplete LU factorization

$\eta_U = \zeta_U = 0$ ▷ $\mathbb{U}$ inverse norm estimators
$\eta_L = \zeta_L = 0$ ▷ $\mathbb{L}$ inverse norm estimators
for all $k \in [1, n]$ **do** ▷ Main loop
  $v_{[k:n]} = (\mathbb{D}_L \mathbb{A} \mathbb{D}_R)_{k,[k,n]}$ ▷ Initialize by kth row
  for all $i \in [1 : k-1] : \mathbb{L}_{ki} \neq 0$ **do**
    $v_{[k:n]} = v_{[k:n]} - \mathbb{L}_{ki}\mathbb{D}_{ii}\mathbb{U}_{i,[k:n]}$
    $v_{[k:n]} = v_{[k:n]} - \mathbb{L}_{ki}\mathbb{D}_{ii}\hat{\mathbb{U}}_{i,[k:n]}$
  end for
  for all $i \in [1 : k-1] : \hat{\mathbb{L}}_{ki} \neq 0$ **do**
    $v_{[k:n]} = v_{[k:n]} - \hat{\mathbb{L}}_{ki}\mathbb{D}_{ii}\mathbb{U}_{i,[k:n]}$
  end for
  $\mathbb{D}_{kk} = \text{sign}(v_k)\max(\tau_p, |v_k|)$ ▷ Pivot modification
  $v_{[k:n]} = v_{[k:n]}/\mathbb{D}_{kk}$
  $\mathbb{C}_{\mathbb{U}} =$ ESTIMATOR$(k, v, \eta_{\mathbb{U}}, \zeta_{\mathbb{U}})$ ▷ See algorithm 6.19
  for all $i \in [k, n] : v_k \neq 0$ **do** ▷ Assemble $\mathbb{U}$ and $\hat{\mathbb{U}}$ kth rows
    if $\mathbb{C}_{\mathbb{U}}|v_k| \geq \tau_1$ **then** $\mathbb{U}_{ki} = v_k$
    else if $\mathbb{C}_{\mathbb{U}}|v_k| \geq \tau_2$ **then** $\hat{\mathbb{U}}_{ki} = v_k$
    end if
  end for
  $v_{[k:n]} = (\mathbb{D}_L \mathbb{A} \mathbb{D}_R)_{[k,n],k}$ ▷ Initialize by kth column
  for all $i \in [1 : k-1] : \mathbb{U}_{ik} \neq 0$ **do**
    $v_{[k:n]} = v_{[k:n]} - \mathbb{U}_{ik}\mathbb{D}_{ii}\mathbb{L}_{[k:n],i}$
    $v_{[k:n]} = v_{[k:n]} - \mathbb{U}_{ik}\mathbb{D}_{ii}\hat{\mathbb{L}}_{[k:n],i}$
  end for
  for all $i \in [1 : k-1] : \hat{\mathbb{U}}_{ik} \neq 0$ **do**
    $v_{[k:n]} = v_{[k:n]} - \hat{\mathbb{U}}_{ik}\mathbb{D}_{ii}\mathbb{L}_{[k:n],i}$
  end for
  $v_{[k:n]} = v_{[k:n]}/\mathbb{D}_{kk}$
  $C_L =$ ESTIMATOR$(k, v, \eta_{\mathbb{L}}, \zeta_{\mathbb{L}})$ ▷ See algorithm 6.19
  for all $i \in [k, n] : v_k \neq 0$ **do** ▷ Assemble $\mathbb{L}$ and $\hat{\mathbb{L}}$ kth columns
    if $C_L|v_k| \geq \tau_1$ **then** $\mathbb{L}_{ik} = v_k$
    else if $C_L|v_k| \geq \tau_2$ **then** $\hat{\mathbb{L}}_{ik} = v_k$
    end if
  end for
end for
for $i \in [k, n]$ **do** ▷ Final $\mathbb{L}$, $\mathbb{D}$, and $\mathbb{U}$ rescaling
  for $j \in [1 : k] : \mathbb{L}_{kj} \neq 0$ **do**
    $\mathbb{L}_{kj} = \mathbb{L}_{kj}\mathbb{D}_{L,kk}/\mathbb{D}_{L,jj}$
  end for
  $\mathbb{D}_{kk} = \mathbb{D}_{kk}/(\mathbb{D}_{L,kk}\mathbb{D}_{R,kk})$
  for $j \in [k : n] : \mathbb{U}_{ij} \neq 0$ **do**
    $\mathbb{U}_{kj} = \mathbb{U}_{kj}\mathbb{D}_{R,kk}/\mathbb{D}_{R,jj}$
  end for
end for

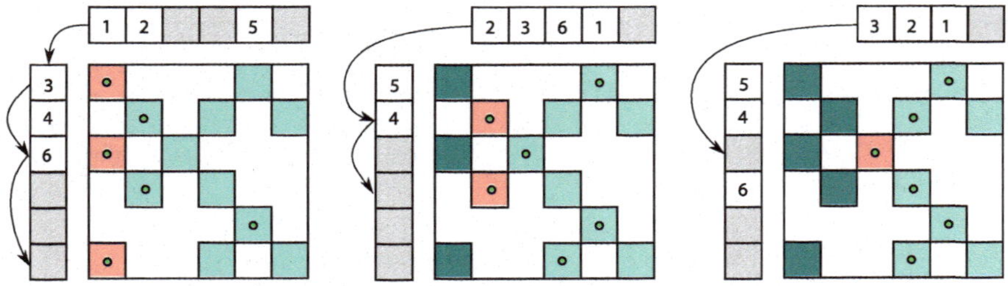

Fig. 6.28 Illustration for traversal of three columns for matrix stored by rows. Teal squares correspond to already considered non-zeros, cyan squares correspond to yet to be considered non-zeros, and pink squares correspond to the elements in the currently considered column. Green dots correspond to positions in I_1 array. Arrays I_2 and I_3 are displayed to the left and top of the matrix, respectively. Gray squares in arrays I_2 and I_3 are NaN values

Algorithm 6.19 Inverse norm estimator

function ESTIMATOR(v, k, η, ζ) ▷ Input
  $\mu_+ = -\eta_k + 1$
  $\mu_- = -\eta_k - 1$
  $s_+ = s_- = 0$
  for all $j \in [k+1:n] : v_j \neq 0$ **do**
    $s_+ = s_+ + |\eta_j + v_j \mu_+|$
    $s_- = s_- + |\eta_j + v_j \mu_-|$
  end for
  $c = \mu_-$
  if $s_+ > s_-$ **then** $c = \mu_+$ **end if**
  $\eta_{[k+1:n]} = \eta_{[k+1:n]} + c v_{[k+1:n]}$ ▷ Update solution η
  $C_1 = \max(|\mu_+|, |\mu_-|)$
  $\mu_+ = -\zeta_k + 1$
  $\mu_- = -\zeta_k - 1$
  $n_+ = n_- = 0$
  for all $j \in [k+1:n] : v_j \neq 0$ **do**
    $v_+ = |\zeta_j + v_j \mu_+|$
    $v_- = |\zeta_j + v_j \mu_-|$
    if $v_+ > \max(2|\zeta_j|, 1/2)$ **then** $n_+ = n_+ + 1$ **end if**
    if $v_- > \max(2|\zeta_j|, 1/2)$ **then** $n_- = n_- + 1$ **end if**
    if $|\zeta_j| > \max(2v_+, 1/2)$ **then** $n_+ = n_+ - 1$ **end if**
    if $|\zeta_j| > \max(2v_-, 1/2)$ **then** $n_- = n_- - 1$ **end if**
  end for
  $c = \mu_-$
  if $n_+ > n_-$ **then** $c = \mu_+$ **end if**
  $\zeta_{[k+1:n]} = \zeta_{[k+1:n]} + c v_{[k+1:n]}$ ▷ Update solution ζ
  $C_2 = \max(|\mu_+|, |\mu_-|)$
  return $\max(C_1, C_2)$ ▷ Output
end function

6.5.4 Multi-Level Factorization

For dynamic pivoting in the Crout LU method, we adopt the multi-level strategy proposed in [36]. If during the Crout LU factorization on level l a small diagonal pivot is encountered or the estimated condition number of the inverse factors $|\mathbb{L}^{-1}|$ and $|\mathbb{U}^{-1}|$ is too large, the computed factorization of the current row and column is abandoned and delayed to the next level. The subsequent elimination is performed without consideration of the delayed rows and columns. When the factorization is completed, the matrix is reordered symmetrically to place all the delayed elements to the end of the matrix. Thus we obtain the following matrix:

$$\mathbb{P}_l \tilde{\mathbb{A}}_l \mathbb{P}_l^T = \begin{bmatrix} \tilde{\mathbb{B}}_l & \tilde{\mathbb{F}}_l \\ \tilde{\mathbb{E}}_l & \tilde{\mathbb{C}}_l \end{bmatrix} = \begin{bmatrix} \tilde{\mathbb{L}}_l \tilde{\mathbb{D}}_l \tilde{\mathbb{U}}_l + \mathcal{E} & \tilde{\mathbb{F}}_l \\ \tilde{\mathbb{E}}_l & \tilde{\mathbb{C}}_l \end{bmatrix}, \tag{6.2}$$

where $\mathbb{P}_l$ denotes the permutation matrix on level l, $\tilde{\mathbb{A}}_l = \mathbb{D}_{L,l}\mathbb{P}_{L,l}\mathbb{A}_l\mathbb{P}_{R,l}\mathbb{D}_{R,l}$ is the rescaled and permuted matrix on level l, and $\mathcal{E}$ is the factorization error. The preprocessing of the matrix is performed on each level. The unscaled $\mathbb{E}_l$ and $\mathbb{F}_l$ matrices are stored by rows and columns, respectively. Note that if $l > 1$, then rows of $\mathbb{E}_i$ and columns of $\mathbb{F}_i$ for $1 \le i < l$ have to be reordered according to permutation in $\mathbb{P}_l$.

The approximate Schur complement is computed for the non-factorized part and is declared to be the next level matrix:

$$\mathbb{A}_{l+1} = \mathbb{S}_l = \tilde{\mathbb{C}}_l - \tilde{\mathbb{E}}_l \tilde{\mathbb{U}}_l^{-1} \tilde{\mathbb{D}}_l^{-1} \tilde{\mathbb{L}}_l^{-1} \tilde{\mathbb{F}}_l. \tag{6.3}$$

Computing Schur complement is performed as follows. We first compute entirely matrix LF $:= \tilde{\mathbb{L}}_l^{-1}\tilde{\mathbb{F}}_l$ by performing the forward substitution with $\tilde{\mathbb{L}}_l^{-1}$ factor on sparse right-hand sides representing columns of $\tilde{\mathbb{F}}_l$. Both first-order and second-order factors of $\mathbb{L}$ are used during the elimination. The operation is performed efficiently due to storage of intermediate results in the ordered dense linked list. Once LF is computed and stored by columns, we reassemble it into the row-wise format. Then we perform the backward substitution with $\tilde{\mathbb{U}}_l$ matrix on rows of $\tilde{\mathbb{E}}_l$ matrix and rescale them by $\tilde{\mathbb{D}}_l^{-1}$. As a result, we obtain EU $:= \tilde{\mathbb{E}}_l\tilde{\mathbb{U}}_l^{-1}\tilde{\mathbb{D}}_l^{-1}$. Further, we construct Schur complement $\mathbb{S}_l$ row by row. To this end, we first put a row of $\mathbb{C}_l$ matrix into an unordered dense linked list structure. Then, we consider column positions of non-zero coefficients of the current row of EU matrix and add to the unordered dense linked list corresponding sparse rows of LF matrix with non-zero coefficients from EU matrix. Thus, we get a row of the Schur complement $\mathbb{S}_l$. Then EU and LF matrices are abandoned.

A dropping strategy is applicable to small non-zero elements in EU, LF, and $\mathbb{S}_l$. After a row of EU or a column of LF is assembled, its norm is computed and all the elements smaller than τ_2 times the norm are dropped. For Schur complement $\mathbb{S}_l$, the multiplication of EU and LF is first performed without assembly of matrix $\mathbb{S}_l$; only norms of rows and columns of $\mathbb{S}_l$ are computed. An element is dropped if it is smaller than τ_2 times the row norm and τ_2 times the column norm. The dropping strategy may be disabled.

The factorization proceeds recursively by setting $\mathbb{A}_{l+1} = \mathbb{S}_l$. The permutation and scaling factors are accumulated. The resulting factors of each level are unscaled but the permutation is retained. The solution process with the multi-level algorithm involves recursive backward and forward substitution on each level. Indeed, given the exact factorization of matrix $\mathbb{A}_l$

$$\mathbb{A}_l = \begin{bmatrix} \mathbb{B}_l & \mathbb{F}_l \\ \mathbb{E}_l & \mathbb{C}_l \end{bmatrix} = \begin{bmatrix} \mathbb{I} & \\ \mathbb{E}_l\mathbb{B}_l^{-1} & \mathbb{I} \end{bmatrix} \begin{bmatrix} \mathbb{B}_l & \\ & \mathbb{S}_l \end{bmatrix} \begin{bmatrix} \mathbb{I} & \mathbb{B}_l^{-1}\mathbb{F}_l \\ & \mathbb{I} \end{bmatrix}, \tag{6.4}$$

with $\mathbb{S}_l = \mathbb{C}_l - \mathbb{E}_l\mathbb{B}_l^{-1}\mathbb{F}_l$, the solution of the following system on level l

$$\mathbb{A}_l \begin{bmatrix} \mathbf{u}_l \\ \mathbf{y}_l \end{bmatrix} = \begin{bmatrix} \mathbf{f}_l \\ \mathbf{g}_l \end{bmatrix}, \tag{6.5}$$

reduces to

$$\mathbf{f} = \mathbb{B}_l^{-1}\mathbf{f}_l, \tag{6.6}$$
$$\mathbf{g} = \mathbf{g}_l - \mathbb{E}_l\mathbf{f}, \tag{6.7}$$
$$\mathbf{y}_l = \mathbb{S}_l^{-1}\mathbf{g}, \tag{6.8}$$
$$\mathbf{u}_l = \mathbf{f} - \mathbb{B}_l^{-1}\mathbb{F}_l\mathbf{y}_l. \tag{6.9}$$

The solution of systems in (6.6), (6.9) with matrix $\mathbb{B}_l = \mathbb{L}_l\mathbb{D}_l\mathbb{U}_l$ involves forward and backward substitutions with already computed factors. The step in (6.8) solves the system (6.5) by (6.6)–(6.9) recursively with matrix $\mathbb{A}_{l+1} = \mathbb{S}_l$ and the right-hand side $\mathbf{g}$ until the last level is reached. Before the solution procedure, the input right-hand side vector is reordered with $\mathbb{P}_R$ permutation matrix and the final solution is reordered with $\mathbb{P}_C$ permutation matrix.

6.5.5 INMOST Linear Solver Routines

Almost all INMOST internal solvers use the BiCGStab(ℓ) iterative method for the linear system solution and the restricted Additive Schwarz method for parallelization. However, the solvers differ in subdomain preconditioners. The following combinations are implemented for internal preconditions:

- *INNER_ILU2*: Matrix is rescaled using norm equilibration and factorized by the row-wise ILU2 method;
- *INNER_MPTILU2*: Matrix is permuted to maximize transversal product and rescaled to equilibrate row and column norms and factorized with the row-wise ILU2 method;

- *INNER_MPTILUC*: Matrix is permuted to maximize transversal product, symmetrically permuted to reduce fill-in, rescaled to equilibrate row and column norms, and factorized with the Crout ILU2 method;
- *INNER_MLMPTILUC*: Nested multi-level factorization strategy is adopted. Matrix on level l is permuted to maximize transversal product, symmetrically permuted to reduce fill-in, rescaled to I-dominant matrix, and factorized with the Crout ILU2 method with delayed pivoting. The Schur complement is computed and declared to be the matrix at level $l + 1$. The multi-level algorithm proceeds until the whole matrix is factored.

These solvers are characterized by the following features. The *INNER_ILU2* is the most efficient choice for easy matrices. However, in complicated cases it may fail to solve a linear system. For instance, in the reservoir simulation, the Jacobian matrix becomes stiffer as the model time step increases, and the solver may fail to converge. This results in undesirable cuts of time steps or the simulation restart with a more robust solver. Fortunately, INMOST offers functionality for automatic optimal parameter tuning using simulated annealing algorithm [28].

The *INNER_MLMPTILUC* method is supposed to be the most robust choice. However, sometimes it may produce large and dense Schur complement matrix despite the dropping of small elements. In this case, the factorization process becomes prohibitively slow. In other cases, dynamic pivoting of *INNER_MLMPTILUC* hits the spot and the method significantly outperforms its simpler analog, *INNER_MPTILUC*.

For most cases, *INNER_MPTILUC* is the method of choice as it provides the most predictable result. If matrices become more stiff during simulation, the method adapts automatically the dropping strategy due to condition number estimation for the inverse triangular factors.

Another INMOST internal solver, K3BIILU2, uses for parallelization the Block Incomplete Inverse LU method [89] instead of the restricted Additive Schwarz method. For internal preconditioning, K3BIILU2 uses Crout ILU2 factorization with preliminary rows and columns norms equilibration. K3BIILU2 features simultaneous MPI and OpenMP parallelization.

The linear solvers are realized with the class *Solver*. The input parameters for the class are the solver name, the solver prefix that selects the parameters from database and the parallel communicator. Apart from the aforementioned internal solvers, a number of external solvers are available to the users, such as Trilinos [82], PETSc [29], and SuperLU or SuperLU_dist [105].

The *Solver* class has two main methods *SetMatrix* and *Solve*. Method *SetMatrix* accepts an object corresponding to a sparse matrix of class *Sparse::Matrix* and computes a preconditioner for the iterative method. Method *Solve* accepts objects corresponding to the right-hand side vector and the solution vector of type *Sparse::Vector*; it invokes the specified method for the system solution and writes the result to the solution vector.

6.6 Automatic Differentiation for Jacobian and Hessian Calculation

INMOST programming platform includes a module for automatic differentiation. The primary objective of the module is to facilitate development of mathematical models that require the matrices of the first derivatives (Jacobian matrix) and the second derivatives (Hessian matrix). Construction of the Jacobians is required for discretizations with the implicit time integration or steady-state problems. The module is exceptionally useful for the development of complex nonlinear models, rf. to Chaps. 3, 4, 5. Open sources of examples of module application are available at INMOST website [7] in "Examples" section.

6.6.1 Basic Structures and Realization Details

The Jacobian (or Hessian) matrix is constructed automatically during arithmetic operations with data. The automatic construction is based on the chain rule. The chain rule declares that despite the complexity of a differentiated expression, the final expression for the derivatives is a sum of partial derivatives with coefficients. Arithmetic operations are declared for the following data types:

- floating-point constants;
- unknowns with single derivative;
- variables with a linear combination of partial derivatives.

The first type of data is the standard C++ type *double*; the last two types are represented by classes. The class representing an *unknown* with single derivative contains the value of the unknown and its position in the Jacobian. This class compactly represents in memory basic unknowns of the model. The class for *variable* with a linear combination of partial derivatives contains a value of the variable and a sparse vector of pairs of values. Each pair of values consists of a coefficient of derivative and its position in Jacobian. This class is used to represent in memory an intermediate result of arithmetic operations on basic unknowns of the model. Each arithmetic operation is represented technically by an expression template class. Each template of the class provides functions to retrieve the value of the expression *GetValue()*, to compute the first-order derivatives *GetJacobian()* or the second-order derivatives *GetHessian()*. The function for retrieval of the derivatives realizes the chain rule for differentiation of composite functions corresponding to the arithmetic operation defined by the class. For example, function *GetJacobian()* for arithmetic operation $\cos(x)$ realizes $-\sin(x)\mathrm{d}x$. Here, $\mathrm{d}x$ is the call to function *GetJacobian()* for argument x with coefficient $-\sin(x)$. For user convenience, the arithmetic operations are constructed using overloaded functions corresponding to class templates, e.g., $\sin(x)$; or using overloaded operators, e.g., operator $+$. The template of the overloaded function $\sin(x)$ returns templated class *sin_expression*. Depending on the class of the argument x,

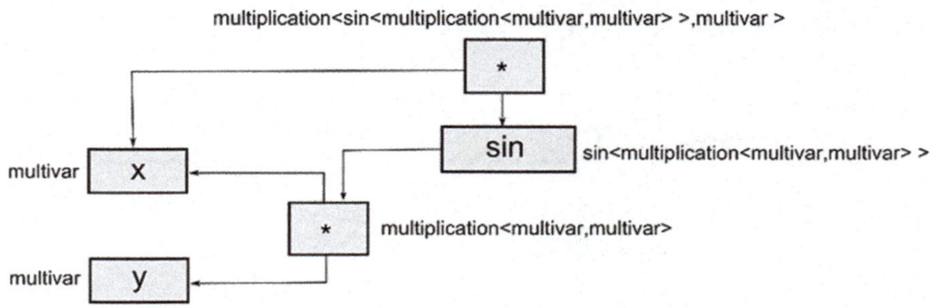

Fig. 6.29 Schematic representation for construction of expression template tree for arithmetic operation $\sin(x \cdot y) \cdot x$

Fig. 6.30 Linear combination of sparse vectors

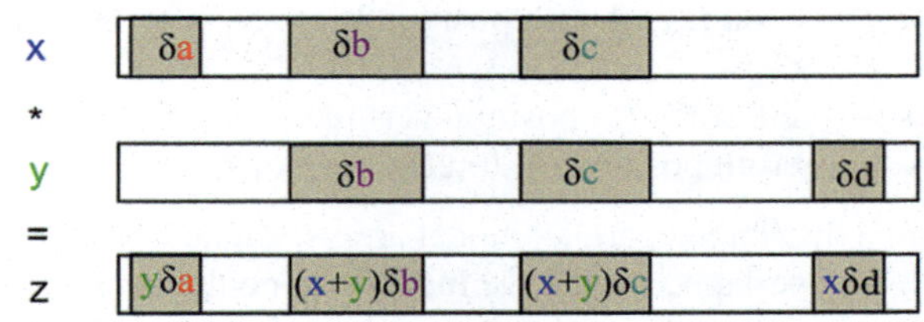

e.g., *double*, the compiler builds the actual function $\sin(x)$ that returns the actual class $sin\_expression < double >$. During compilation, a tree that is based on class templates is built (see Fig. 6.29) and the function *GetJacobian()* realizes depth tree traversal for assembly of the sparse vector of the derivatives and their coefficients. To accelerate the computation, each class stores the result of corresponding arithmetic operation and in some cases the derivative multipliers, in order to avoid depth tree traversal for function *GetValue()*.

The construction is rather flexible: it can realize even such arithmetic operation as interpolations within tabulated data. Let us consider operation $z = x \cdot y$, where x contains the partial derivatives $\partial a + \partial b + \partial c$ and y contains the partial derivatives $\partial b + \partial c + \partial d$. The result of function *GetJacobian()* for such expression is the sparse vector of the partial derivatives $y\partial a + (x + y)\partial b + (x + y)\partial c + x\partial d$. For computation of the expression, we need operations for multiplication of a sparse vector by a coefficient and addition of sparse vectors. Such operation is called linear combination of sparse vectors (Fig. 6.30).

The same operation is used for computation of the incomplete LU factorization to precondition the iterative solution of large sparse systems. A computationally efficient structure for such operation is a dense linked list. For automatic differentiation, we use an unordered dense linked list that has $O(1)$ complexity for insertion of new elements, see Fig. 6.31. The list is composed of two dense arrays: array of values A and array of the next non-zero element J. By default, array J is filled with EOR value that corresponds to the largest possible integer value. EOR corresponds to the absence of non-zero element at current position. The zeroth position of J contains the

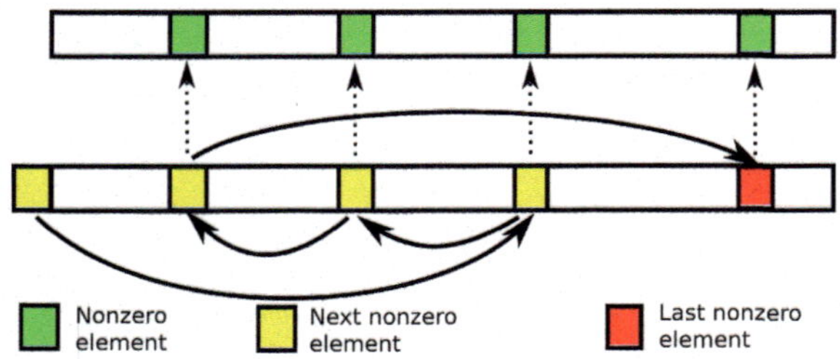

Fig. 6.31 Unordered dense linked list structure

position of the last inserted non-zero element. Insertion of a new non-zero element into position i is performed as follows:

- EOR value at the ith position is replaced by the value at the zeroth position of J array;
- A value at the ith position gets the value of the derivative coefficient;
- the zeroth position of J gets the value i.

In a distributed parallel model arrays, A and J have the size equal to the size of the local overlapped matrices. Indices for nonlocal unknowns are mapped to the end of extended arrays J and A. In a shared parallel model, each processor keeps a copy of the dense linked list.

Thus in operation $z = x \cdot y$ function *GetJacobian()* is called automatically for a class tree corresponding to operation $x \cdot y$ and the result is assigned to z.

Using the expression template classes, we also implemented an abstraction for lazy evaluation of operations. Such abstraction represents a tree of classes, which evaluates by producing the aforementioned tree of classes. The latter corresponds to stored arithmetic operations upon substitution of the mesh element. Such technology allows the user to realize functions based on automatic differentiation for which the expression for the argument is not known beforehand and cannot be defined by templated function. For example, in reservoir simulation, one needs to compute the pressure gradient and sometimes one needs to account for the gradient of the capillary pressure. In both cases, the gradient is computed in the same way; however, the argument is not known beforehand. Usually, the capillary pressure is a tabulated function of fluid saturation; such expression should be properly incorporated into the gradient calculation. The technology includes description of conditional operations and operations describing the stencil of the spatial differential operator. Hence, the technology completely separates model description and numerical computations. Such feature is very useful for future transfer of numerical computation of a nonlinear residual and its Jacobian onto acceleration units (for example, GPU) without necessity for the user to code on dedicated programming language. By analyzing the stored expression, the CPU can build the accelerating unit code that executes the expression, prepares, and feeds all data needed for the execution.

6.6.2 Interfaces for Automatic Differentiation

The basis for the module interface is composed of

- a C++ class *unknown* to represent a basic unknown of the system;
- a C++ class *variable* to represent a value of function with its first-order partial derivatives;
- a C++ class *hessian_variable* to represent a value of function with both its first-order and second-order partial derivatives;
- a set of C++ overloaded operators for construction of arbitrary expressions based on expression templates that form a templated tree of classes corresponding to operations on variables;
- a C++ class *stored_variable_expression* that can store expression template and evaluate it on demand;
- a (hidden to the user) dense linked list structure *Sparse::RowMerger* is used for fast addition of sparse vectors corresponding to partial derivatives.

To accumulate intermediate results, one may associate data, representing values with sparse vectors of partial derivatives, with elements of the grid. The stored data can be directly used in further arithmetic operations. The storage of the data in files of ".pmf" and ".xml" formats as well as interchange of data between processors is also supported.

INMOST implements a templated class *Matrix* of the dense linear algebra module that provides operations on dense matrices of values with specified type. Many *blas* and *lapack* operations are implemented in this class: sub-matrix access, addition, subtraction, multiplication, system solution, matrix inversion, singular value decomposition, pseudo-solve and pseudo-inverse, and so on. The whole functionality is supported for matrices of values with the first and second partial derivatives. A set of simple rules prescribes the outcome when two matrices with different types of elements are involved in operations, *e.g.,* multiplication of a matrix of *doubles* with a matrix of *variables* results in a matrix of *variables*. This functionality is heavily used in the blood coagulation model, see Chap. 5. The following additional interfaces are available:

- a C++ class $Matrix < T >$ where T can be *double*, *unknown*, *variable* or *hessian _variable*;
- a set of C++ overloaded operators for operations on $Matrix < T >$;
- a C++ class *stored_block_variable_expression* that can store operations on matrices and evaluate them on demand.

To connect mesh data and automatic differentiation, we introduce an *Automatizator* class. The class *Automatizator* controls main unknowns of the model, their activation, and enumeration. This class also regulates the dense linked list size, which is used to speed up computation of linear combinations of sparse vectors.

The user supplies mesh data *Tag* corresponding to main unknowns on mesh elements to the *Automatizator* class via function *RegisterTag*. Calling the function *EnumerateEntries* causes the enumeration of main unknowns and allocation of memory for the support of data structures. The unknowns may be distributed among multiple meshes. The user can use the class *dynamic_variable* with the index returned by *RegisterTag* to retrieve the *unknown* on each element of the mesh.

A more advanced block-structured organization of unknowns is possible. An *AbstractEntry* sub-class allows to group the mesh data into blocks of unknowns and register them with the object of the *Automatizator* class via function *RegisterEntry*.

The main functionality of the "*AbstractEntry*" class covers the following:

- *Value* provides a matrix of values of unknowns on mesh element;
- *Index* provides a matrix of indices of unknowns on mesh element; these indices are positions of unknowns in the Jacobian matrix;
- *Unknown* provides a matrix of elements with type *unknown* on mesh element;
- *MatrixSize* determines the size of the returned matrix.

Each *value*, *index*, and *unknown* can be accessed individually if it is undesirable to get the whole matrix; the assembly of the matrix is being avoided in this case.

Various scenarios of unknown organization are covered by sub-classes with extended functionality:

- *SingleEntry* to represent a single unknown on mesh elements, used within *RegisterTag* method;
- *VectorEntry* to represent multiple unknowns on mesh elements (possibly with variable length);
- *BlockEntry* to represent a fixed-size block of unknowns on mesh elements;
- *StatusBlockEntry* allows to change the status of unknowns to constants in the block;
- *MultiEntry* is a union of entries of different types, *e.g.*, an object of "*SingleEntry*" can be used together with an object of "*VectorEntry*".

Each of these sub-classes is inherited from the *AbstractEntry* class and retains the original functionality.

Within the *Automatizator*, an index data is created and associated to each entry of a block of unknowns, and enumeration for unknowns is performed on every mesh element containing the data related to the unknowns. Each block is enumerated consequently on each element of the mesh. This leads to the block-structured organization of the Jacobian matrix. Based on this functionality, one can split the Jacobian matrix into blocks of physical processes. Potentially, the information of the block structure of the matrix can be shared with linear solvers.

To facilitate assembly of the residual vector and associated matrices, we introduced a *Residual* class. The class *Residual* contains a vector corresponding to residual and a sparse matrix corresponding to Jacobian. By performing operations on the *Residual* class, the user directly changes structures of the type *Sparse::Vector* and *Sparse::Matrix*, which are later supplied to the linear solution method. During the assembly stage, a sparse matrix is stored in INMOST as a set of sparse vectors corresponding to rows of the matrix, i.e., class *Sparse::Row*. Each sparse vector is expandable and allows fast modifications. The same vector is used to store partial derivatives in the type *variable* of the automatic differentiation module. When an object of the class *Residual* is accessed by index (similar to access to array R via square brackets $R[i]$), an object of the class $variable\_reference$ with references to corresponding entry in the residual vector and the row in the sparse matrix is returned. The object of this class can enter expression templates from the automatic differentiation module and thus admits the same operations as an object of the *variable* type. Assignment to an object of this class results in modification of the sparse matrix and the residual vector stored in the object of the *Residual* class.

On top of that, an object of the class *Residual* can be accessed by a matrix of indices, (i.e., $R[\mathcal{I}]$ where $\mathcal{I}$ is the matrix of indices of type $Matrix < int >$); then it returns a matrix with the type $Matrix < variable\_reference >$. On the assignment, the underlying sparse matrix and the residual vector are modified in a block-structured fashion. In the future, we shall introduce block-structured sparse matrices and use them in the *Residual* class to take the advantage of the fast block-structured assembly.

The presented functionality provides seamless integration between the automatic differentiation module and the sparse linear algebra module. The functionality is readily available for distributed and shared parallel computations.

6.7 Nonlinear Solvers

A set of methods that address the iterative solution of nonlinear systems are implemented using INMOST modules.

Most of the nonlinear systems appearing in the previous chapters were solved by the Newton method. However, the Newton method requires a good initial guess and a smooth enough (i.e., with Lipschitz continuous derivative) function to converge. The Newton iterations may easily diverge or dangle around the true solution for a large class of problems involving non-differentiable functions (i.e., modulus operation, upstream-differencing) or functions with inflection points. Such issues are encountered when one applies for the approximate solution of the convection–diffusion equation the nonlinear FV discretization satisfying the discrete maximum principle, rf. to Chap. 2.

In the following, we consider a set of practical methods for the solution of nonlinear problems.

6.7.1 Newton Method

Let $\mathbf{x}_k \in \mathbb{R}^N$ represent the vector of N unknowns on the kth Newton iteration and $\mathcal{R}_k = \mathcal{R}(\mathbf{x}_k) \in \mathbb{R}^N$ represent the corresponding residual vector. The Jacobian of $\mathcal{R}_k$ is $\mathcal{J}_k = \partial\mathcal{R}(\mathbf{x}_k)/\partial\mathbf{x}_k \in \mathbb{R}^{N\times N}$. In unsteady simulations, the initial guess $\mathbf{x}_0$ can be taken from the previous time step. In steady simulations, the choice of the initial guess may be an issue.

If the L_2-norm $\|\mathcal{R}_k\|_{L_2}$ of the residual satisfies the convergence criterion, we declare $\mathbf{x}_k$ as the solution of the nonlinear system. Otherwise, the system for the Newton update is solved:

$$\mathcal{J}_k \Delta \mathbf{x}_k = -\mathcal{R}_k, \quad \mathbf{x}_{k+1} = \mathbf{x}_k + \Delta \mathbf{x}_k, \tag{6.10}$$

then the solution proceeds with consideration of the residual for $\mathbf{x}_{k+1}$.

Implementation of the Newton method is easy within the INMOST platform. Given an assembly procedure for the residual $\mathcal{R}_k$ via *Residual* class, INMOST provides assembly for the sparse Jacobian with the automatic differentiation module. The system (6.10) is efficiently solved with available linear solvers for systems with distributed sparse matrices.

There are two possible ways for improvement of the Newton method. Introduction of parameter α into the update procedure, $\mathbf{x}_{k+1} = \mathbf{x}_k + \alpha \Delta \mathbf{x}_k$ leads to backtracking and line-search methods. Inexact computation of $\mathcal{J}_k$ leads to a family of Quasi-Newton methods including Picard iteration, Broyden update, and trust-region methods.

6.7.2 Line-Search Methods

Let us assume that the update $\Delta \mathbf{x}_k$ for the kth nonlinear iteration was provided by the Newton method. The line-search method consists in finding such a parameter α that $\|\mathcal{R}(\mathbf{x}_k + \alpha \Delta \mathbf{x}_k)\|$ is minimal and declares $\mathbf{x}_{k+1} = \mathbf{x}_k + \alpha \Delta \mathbf{x}_k$ as the next iterative guess. The method does not change the direction of the provided update $\Delta \mathbf{x}_k$.

Usually, interval $\alpha \in [0, 1]$ is considered. In this case, a simple searching method based on golden section ratio can be used. Line-search method is very useful if the computation of the residual is considerably less expensive than assembly of the Jacobian. Otherwise, it may be more beneficial to perform another Newton step in another direction.

6.7.3 Anderson Acceleration Method

Let $\mathcal{R}(\mathbf{x}) \in \mathbb{R}^N$ be a residual for a nonlinear system of equations whose evaluation is expensive. Let $\mathbf{x}_{k-m}, \ldots, \mathbf{x}_k \in \mathbb{R}^N$ represent last m iterates with $\mathcal{R}_{k-m}, \ldots, \mathcal{R}_k \in \mathbb{R}^N$ corresponding residuals, $\mathcal{R}_i = \mathcal{R}(\mathbf{x}_i)$. Then, the estimate for the next iteration $\mathbf{x}_{k+1}^{AA}$ using the Anderson acceleration method is computed as a convex linear combination with coefficients α_j and mixing parameter γ [23]:

$$\mathbf{x}_{k+1}^{AA} = \sum_{j=k-m}^{k} \alpha_j \mathbf{x}_j + \gamma \tilde{\mathcal{R}}_{k+1}, \quad \tilde{\mathcal{R}}_{k+1} = \sum_{j=k-m}^{k} \alpha_j \mathcal{R}_j, \quad \left\| \tilde{\mathcal{R}}_{k+1} \right\| \to \min, \quad \sum_{j=k-m}^{k} \alpha_j = 1. \tag{6.11}$$

The minimization problem for vector of coefficients $\boldsymbol{\alpha} = \left[\alpha_{k-m} \ldots \alpha_k\right]^T \in \mathbb{R}^m$ is solved by considering an $(m+1) \times (m+1)$ saddle-point system:

$$\begin{bmatrix} \mathcal{R}^T \mathcal{R} & \beta \mathbf{1} \\ \beta \mathbf{1}^T & 0 \end{bmatrix} \begin{bmatrix} \boldsymbol{\alpha} \\ \lambda \end{bmatrix} = \begin{bmatrix} \mathbf{0} \\ \beta \end{bmatrix}, \tag{6.12}$$

where $\mathcal{R} = \left[\mathcal{R}_{k-m} \ldots \mathcal{R}_k\right] \in \mathbb{R}^{N \times m}$ and $\beta = \mathrm{trace}(\mathcal{R}^T \mathcal{R})$.

Observing that the kth step Newton update $\|\Delta \mathbf{x}_k\|$ converges to zero as the method converges to the solution, one may optimize for $\boldsymbol{\alpha}$ by considering a set of Newton updates instead of the residuals [109], i.e., solve

$$\left\| \sum_{j=k-m}^{k} \alpha_j \Delta \mathbf{x}_j \right\| \to \min, \tag{6.13}$$

and the next iterate $\mathbf{x}_{k+1}^{AA}$ is defined by

$$\mathbf{x}_{k+1}^{AA} = \gamma \sum_{j=k-m}^{k} \alpha_j \mathbf{x}_j + (1-\gamma) \sum_{j=k-m}^{k} \alpha_j \mathbf{x}_{j+1}^{N} = \sum_{j=k-m}^{k} \alpha_j \left(\mathbf{x}_j + (1-\gamma)\Delta \mathbf{x}_j\right), \tag{6.14}$$

where $\mathbf{x}_{k+1}^{N} = \mathbf{x}_k + \Delta \mathbf{x}_k$ is the solution suggested by the Newton method.

Usually, the Anderson acceleration method is not beneficial for the quadratically convergent Newton method [64]. However, it can improve significantly stalling or diverging Newton's method. Also, it is also suitable for Quasi-Newton methods.

The method changes the direction of the Newton update. The line-search method may improve the Anderson acceleration. The efficient strategy is to perform line search between solution provided by the Newton method $\mathbf{x}_{k+1}^{N}$ and the solution accelerated by the Anderson method $\mathbf{x}_{k+1}^{AA}$ as proposed in [138]. Then the next solution is defined as $\mathbf{x}_{k+1} = (1-\eta)\mathbf{x}_k^N + \eta \mathbf{x}_k^{AA}$, where parameter η is determined based on the line-search solution of minimization problem $\left\|\mathcal{R}\left((1-\eta)\mathbf{x}_k^N + \eta \mathbf{x}_k^{AA}\right)\right\| \to \min$.

6.7.4 Halley Method

Given the Hessian matrix $\mathcal{H}_k = \partial^2 \mathcal{R}(\mathbf{x}_k)/\partial \mathbf{x}_k^2$ on the kth iteration, one can solve the nonlinear problem with the two-step Halley method:

$$\mathcal{J}_k \Delta \mathbf{x}_1 = -\mathcal{R}_k, \quad \left(\mathcal{J}_k + \frac{1}{2}\mathcal{H}_k \Delta \mathbf{x}_1\right) \Delta \mathbf{x}_2 = -\mathcal{R}_k, \quad \mathbf{x}_{k+1} = \mathbf{x}_k + \Delta \mathbf{x}_2. \tag{6.15}$$

Compared to the Newton method, it requires an extra step which involves multiplication with the Hessian matrix. Assembly of the Hessian matrix proves to be prohibitively expensive for large problems. However, one may find it useful for small local problems.

6.8 Multi-Physics Model Assembly

INMOST provides the module for assembly of a complex multi-physics model. The idea of the module is to provide basic functionality that allows one to split the problem into physical processes. Each physical process is represented by a sub-model. Each sub-model is responsible for the introduction of unknowns and assembly of a residual associated with a sub-model physical process. The sub-models are coupled and solved together as single model. Coupling between sub-models is organized by the introduction of coupling terms that depend on unknowns in adjacent models.

Let us consider the Jacobian matrix of such multi-physics model assembled of multiple sub-models as illustrated in Fig. 6.32. Each sub-model represents diagonal blocks of the Jacobian, and coupling between sub-models completes off-diagonal terms.

Coupling between two physical processes introduces coupling terms into equations involving unknowns of both processes which have to be accessed. Such couplings can involve dependence of model parameters on unknowns (i.e., porosity

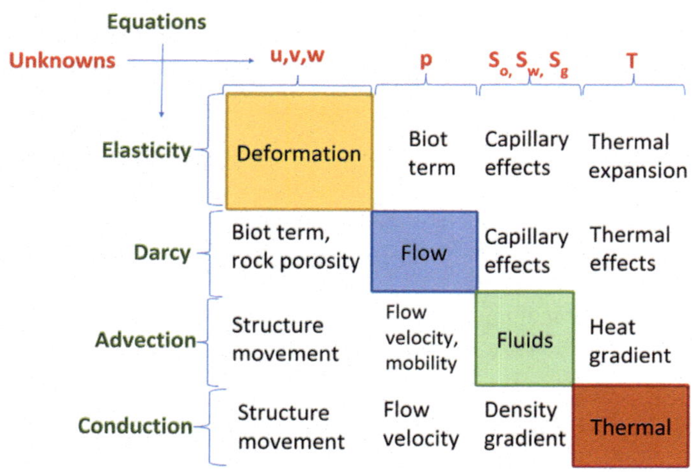

Fig. 6.32 Example of multi-physics model organization for thermo-poro-elasticity problem. The matrix of the sub-models and coupling terms corresponds to the Jacobian matrix. The primary processes (colored boxes) form the diagonal blocks of the Jacobian. The couplings between the models form the off-diagonal parts

depends not only on fluid pressure but on mechanical deformations and temperature); involve additional terms with partial derivatives (i.e., the fluid convection with the heat gradient); or modify the right-hand side (i.e., a set of kinetic reactions). The interaction between the processes can be quite complex and should be discretized accordingly with stability issues in mind. Typically, stability issues are related to the signs of the real part of the eigenvalues in the Jacobian and can be diagnosed by large off-diagonal terms. A well-known issue in discretization of coupled problems is inf-sup stability issue. The inf-sup stability issues were previously addressed in sections 2.6, 2.7 by considering discretizations with matrix coefficients with positive eigenvalues. Modifications to the set of reactions that allow for stable integration with large time steps were considered in section 5.2.

Furthermore, a set of additional constraints should be applied to the unknowns, i.e., fluid saturation cannot be less than zero or above unity, pressure has to satisfy the discrete maximum principle, and so on. In some cases, the value of the unknown defying the constraint can be simply clamped; in other cases, the update has to be backtracked for the sake of local conservation.

Sub-models in INMOST are represented by *AbstractSubModel* class. The programmer has to inherit from this class and implement the following functions:

- *PrepareEntries* introduces unknowns of a process to the model;
- *FillResidual* computes the residual for a process;
- *UpdateMultiplier* performs backtracking of an update to meet constraints;
- *UpdateSolution* updates unknowns during nonlinear iterations;
- *UpdateTimeStep* proceeds to the next time step;
- *AdjustTimeStep* computes an optimal time step for a process;
- *RestoreTimeStep* returns the previous solution in case of nonlinear solver failure.

A set of objects inherited from *AbstractSubModel* is managed by an object from a *Model* class. It also incorporates an object of the *Automatizator* class and named arrays of meshes (class *Mesh*), entries of unknowns (class *AbstractEntry*), submodels (class *AbstractSubModel*), and couplings.

The couplings are represented by *AbstractCoupling* that has functions *PrepareEntries* and *FillResidual*. This module uses heavily all the previously introduced modules. A nonlinear solver that guides the convergence of the multi-physics model to the solution is implemented on top of the *Model* class using the provided functionality.

References

1. Ansys Fluent – Computational fluid dynamics tool. https://www.ansys.com/products/fluids/ansys-fluent.
2. COMSOL Multiphysics – Cross-platform finite element analysis, solver and multiphysics simulation software. https://www.comsol.com/.
3. COOLFluiD – Computational Object-Oriented Libraries for Fluid Dynamics. https://github.com/andrealani/COOLFluiD/.
4. DUNE – the Distributed and Unified Numerics Environment. https://dune-project.org/.
5. Elmer FEM – Open source multiphysical simulation software. https://www.csc.fi/web/elmer.
6. INM RAS cluster. http://cluster2.inm.ras.ru/.
7. INMOST – A toolkit for distributed mathematical modeling. http://www.inmost.org/.
8. Norne – The full Norne benchmark case, a real field black-oil model for an oil field in the Norwegian Sea. https://opm-project.org/?page_id=559.
9. OOFEM – Object Oriented Finite Element Solver. http://www.oofem.org/.
10. OpenFOAM – The Open Source Computational Fluid Dynamics (CFD) Toolbox. https://www.openfoam.com/.
11. OPM – The Open Porous Media initiative. https://opm-project.org/.
12. PETSc – Portable Extensible Toolkit for Scientific Computation. https://www.mcs.anl.gov/petsc/.
13. STAR-CCM+ – CFD-focused multiphysics simulation. https://www.plm.automation.siemens.com/global/en/products/simcenter/STAR-CCM.html.
14. SU2 – Multiphysics Simulation and Design Software. https://su2code.github.io/.
15. Trilinos – Platform for the solution of large-scale, complex multi-physics engineering and scientific problems. http://trilinos.org/.
16. Aavatsmark, I., Barkve, T., Bøe, O., & Mannseth, T. (1998). Discretization on unstructured grids for inhomogeneous, anisotropic media. part I: Derivation of the methods. *SIAM Journal on Scientific Computing*, *19*(5), 1700–1716.
17. Aavatsmark, I., Eigestad, G., Mallison, B., & Nordbotten, J. (2008). A compact multipoint flux approximation method with improved robustness. *Numerical Methods for Partial Differential Equations*, *24*(5), 1329–1360.
18. Ackerer, P., Younes, A., & Mose, R. (1999). Modeling variable density flow and solute transport in porous medium: 1. Numerical model and verification. *Transport in Porous Media*, *35*(3), 345–373.
19. Agélas, L., Eymard, R., & Herbin, R. (2009). A nine-point finite volume scheme for the simulation of diffusion in heterogeneous media. *Comptes Rendus Mathematique*, *347*(11–12), 673–676.

Y. Vassilevski et al., *Parallel Finite Volume Computation on General Meshes*,
https://doi.org/10.1007/978-3-030-47232-0

20. Aliaga, J. I., Bollhöfer, M., Martín, A. F., & Quintana-Ortí, E. S. (2008). Design, tuning and evaluation of parallel multilevel ILU preconditioners. *International Conference on High Performance Computing for Computational Science* (pp. 314–327). Berlin: Springer.
21. Amestoy, P. R., Duff, I. S., Koster, J., & L'Excellent, J.-Y. (2001). A fully asynchronous multifrontal solver using distributed dynamic scheduling. *SIAM Journal on Matrix Analysis and Applications*, *23*(1), 15–41.
22. Amestoy, P. R., Guermouche, A., L'Excellent, J.-Y., & Pralet, S. (2006). Hybrid scheduling for the parallel solution of linear systems. *Parallel Computing*, *32*(2), 136–156.
23. Anderson, D. G. (1965). Iterative procedures for nonlinear integral equations. *Journal of the ACM (JACM)*, *12*(4), 547–560.
24. Andersson, K., Grundfeld, G., Hodgkinson, D., Lindbom, B., & Jackson, C. (1986). *Hydrocoin level 1 final report: verification of groundwater flow models*. Swedish Nuclear Power Inspectorate (SKI): Stockholm.
25. Anuprienko, D., & Kapyrin, I. (2018). Modeling groundwater flow in unconfined conditions: numerical model and solvers' efficiency. *Lobachevskii Journal of Mathematics*, *39*(7), 867–873.
26. Aziz, K., & Settari, A. (1979). *Petroleum reservoir simulation*. London: Applied Science Publishers.
27. Babur, Ö., Smilauer, V., Verhoeff, T., & van den Brand, M. (2015). A survey of open source multiphysics frameworks in engineering. *Procedia Computer Science*, *51*, 1088–1097.
28. Bagaev, D., Konshin, I., & Nikitin, K. (2017). Dynamic optimization of linear solver parameters in mathematical modelling of unsteady processes. *Russian supercomputing days* (pp. 54–66). Berlin: Springer.
29. Balay, S., Abhyankar, S., Adams, M., Brown, J., Brune, P., Buschelman, K., et al. (2019). *Petsc users manual*.
30. Barragy, E., & Carey, G. (1997). Stream function-vorticity driven cavity solution using P finite elements. *Computers & Fluids*, *26*(5), 453–468.
31. Bear, J., & Cheng, A. H.-D. (2010). *Modeling groundwater flow and contaminant transport* (Vol. 23). Berlin: Springer Science & Business Media.
32. Berman, A., & Plemmons, R. (1979). *Nonnegative matrices in the mathematical sciences* (p. 334). New York: Academic Press.
33. Bertolazzi, E., & Manzini, G. (2005). A second-order maximum principle preserving finite volume method for steady convection-diffusion problems. *SIAM Journal of Numerical Analysis*, *43*(5), 2172–2199.
34. Boldyrev, K., Kapyrin, I., Konstantinova, L., & Zakharova, E. (2016). Simulation of strontium sorption onto rocks at high concentrations of sodium nitrate in the solution. *Radiochemistry*, *58*(3), 243–251.
35. Bollhöfer, M. (2001). A robust ILU with pivoting based on monitoring the growth of the inverse factors. *Linear Algebra and its Applications*, *338*(1–3), 201–218.
36. Bollhöfer, M., & Saad, Y. (2006). Multilevel preconditioners constructed from inverse-based ILUs. *SIAM Journal on Scientific Computing*, *27*(5), 1627–1650.
37. Boman, E. G., Çatalyürek, Ü. V., Chevalier, C., & Devine, K. D. (2012). The Zoltan and Isorropia parallel toolkits for combinatorial scientific computing: Partitioning, ordering and coloring. *Scientific Programming*, *20*(2), 129–150.
38. Bouchnita, A. (2017). *Mathematical modelling of blood coagulation and thrombus formation under flow in normal and pathological conditions*. PhD thesis, Universit'e Lyon 1 - Claude Bernard; Ecole Mohammadia d'Ing'enieurs - Universit'e Mohammed V de Rabat - Maroc.
39. Bouchnita, A., Terekhov, K., Nony, P., Vassilevski, Y., & Volpert, V. (2020). A mathematical model to quantify the effects of platelet count, shear rate, and injury size on the initiation of blood coagulation under venous flow conditions. *To appear in Plos One*.
40. Braack, M., & Richter, T. (2006). Solutions of 3D Navier-Stokes benchmark problems with adaptive finite elements. *Computers & Fluids*, *35*(4), 372–392.
41. Brooks, A., & Hughes, T. (1982). Streamline upwind/Petrov-Galerkin formulations for convection dominated flows with particular emphasis on the incompressible Navier-Stokes equations. *Computer Methods in Applied Mechanics and Engineering*, *32*(1–3), 199–259.

42. Brooks, R. H., & Corey, A. T. (1966). Properties of porous media affecting fluid flow. *Journal of the Irrigation and Drainage Division*, *92*(2), 61–90.
43. Bunch, J. R., & Kaufman, L. (1977). Some stable methods for calculating inertia and solving symmetric linear systems. *Mathematics of Computation*, 163–179.
44. Celia, M. A., Bouloutas, E. T., & Zarba, R. L. (1990). A general mass-conservative numerical solution for the unsaturated flow equation. *Water Resources Research*, *26*(7), 1483–1496.
45. Charlton, S. R., & Parkhurst, D. L. (2011). Modules based on the geochemical model PHREEQC for use in scripting and programming languages. *Computers & Geosciences*, *37*(10), 1653–1663.
46. Charny, I. A. (1951). A rigorous derivation of Dupuit's formula for unconfined seepage with seepage surface. *Doklady Akademii Nauk SSSR*, *79*, 937–940.
47. Chen, Z., Huan, G., & Ma, Y. (2006). *Computational methods for multiphase flows in porous media*. Philadelphia: SIAM.
48. Chernyshenko, A., & Vassilevski, Y. (2014). A finite volume scheme with the discrete maximum principle for diffusion equations on polyhedral meshes. *Finite Volumes for Complex Applications VII-Methods and Theoretical Aspects*, 197–205.
49. Coon, E. T., Moulton, J. D., & Painter, S. L. (2016). Managing complexity in simulations of land surface and near-surface processes. *Environmental Modelling & Software*, *78*, 134–149.
50. Cuthill, E., & McKee, J. (1969). Reducing the bandwidth of sparse symmetric matrices. In *Proceedings of the 1969 24th National Conference* (pp. 157–172).
51. Danilov, A., & Vassilevski, Y. (2009). A monotone nonlinear finite volume method for diffusion equations on conformal polyhedral meshes. *Russian Journal of Numerical Analysis and Mathematical Modelling*, *24*(3), 207–227.
52. Davis, T. A. (2011). UMFPACK user guide. *Engineering*, 1–140.
53. Diersch, H. (1998). Treatment of free surfaces in 2D and 3D groundwater modeling. *Mathematische Geologie*, *2*, 17–43.
54. Diersch, H.-J., & Perrochet, P. (1999). On the primary variable switching technique for simulating unsaturated-saturated flows. *Advances in Water Resources*, *23*(3), 271–301.
55. Diersch, H.-J. G. (2013). *FEFLOW: Finite element modeling of flow, mass and heat transport in porous and fractured media*. Berlin: Springer Science & Business Media.
56. Ding, Y., & Jeannin, L. (2001). A new methodology for singularity modelling in flow simulations in reservoir engineering. *Computational Geosciences*, *5*, 93–119.
57. Dotlić, M., Vidović, D., Pokorni, B., Pušić, M., & Dimkić, M. (2016). Second-order accurate finite volume method for well-driven flows. *Journal of Computational Physics*, *307*, 460–475.
58. Droniou, J. (2014). Finite volume schemes for diffusion equations: introduction to and review of modern methods. *Mathematical Models and Methods in Applied Sciences*, *24*(08), 1575–1619.
59. Droniou, J., & Potier, C. L. (2011). Construction and convergence study of schemes preserving the elliptic local maximum principle. *SIAM Journal on Numerical Analysis*, *49*(2), 459–490.
60. Duff, I. S., & Koster, J. (1999). The design and use of algorithms for permuting large entries to the diagonal of sparse matrices. *SIAM Journal on Matrix Analysis and Applications*, *20*(4), 889–901.
61. Elder, J. (1967). Transient convection in a porous medium. *Journal of Fluid Mechanics*, *27*(3), 609–623.
62. Elder, J. W. (1967). Steady free convection in a porous medium heated from below. *Journal of Fluid Mechanics*, *27*(1), 29–48.
63. Elder, J. W., Simmons, C. T., Diersch, H.-J., Frolkovič, P., Holzbecher, E., & Johannsen, K. (2017). The Elder Problem. *Fluids*, *2*(1), 11.
64. Evans, C., Pollock, S., Rebholz, L. G., & Xiao, M. (2020). A proof that Anderson acceleration improves the convergence rate in linearly converging fixed-point methods (but not in those converging quadratically). *SIAM Journal on Numerical Analysis*, *58*(1), 788–810.
65. Eymard, R., Gallouët, T., & Herbin, R. (2000). Finite volume methods. *Handbook of numerical analysis*, *7*, 713–1018.

66. Flemisch, B., Darcis, M., Erbertseder, K., Faigle, B., Lauser, A., Mosthaf, K., Müthing, S., Nuske, P., Tatomir, A., Wolff, M., & Helmig, R. (2011). Dumux: Dune for multi-phase, component, scale, physics,... flow and transport in porous media. *Advances in Water Resources*, *34*(9), 1102–1112.
67. Forsyth, P. A., Wu, Y., & Pruess, K. (1995). Robust numerical methods for saturated-unsaturated flow with dry initial conditions in heterogeneous media. *Advances in Water Resources*, *18*(1), 25–38.
68. Freundlich, H. (1907). Über die adsorption in lösungen. *Zeitschrift für physikalische Chemie*, *57*(1), 385–470.
69. Frolkovič, P., & De Schepper, H. (2000). Numerical modelling of convection dominated transport coupled with density driven flow in porous media. *Advances in Water Resources*, *24*(1), 63–72.
70. Geuzaine, C., & Remacle, J.-F. (2009). Gmsh: A 3-D finite element mesh generator with built-in pre-and post-processing facilities. *International Journal for Numerical Methods in Engineering*, *79*(11), 1309–1331.
71. Ghia, U., Ghia, K. N., & Shin, C. (1982). High-Re solutions for incompressible flow using the Navier-Stokes equations and a multigrid method. *Journal of Computational Physics*, *48*(3), 387–411.
72. Grigorev, F., Plenkin, A., & Kapyrin, I. (2018). To the necessity of taking into account the disposal's structure whilst far field waste input modeling. *Radioactive Waste*, *3*(4), 95–101.
73. Grigor'ev, F. V., Kapyrin, I. V., & Vassilevski, Y. V. (2017). Modeling of thermal convection in porous media with volumetric heat source using the GeRa code (in Russian). *Chebyshevskii Sbornik*, *18*(3), 235–254.
74. Guo, W., & Langevin, C. D. (2002). User's guide to SEAWAT; a computer program for simulation of three-dimensional variable-density ground-water flow. *Techniques of Water-Resources Investigations 06-A7*.
75. Hadgu, T., Karra, S., Kalinina, E., Makedonska, N., Hyman, J. D., Klise, K., et al. (2017). A comparative study of discrete fracture network and equivalent continuum models for simulating flow and transport in the far field of a hypothetical nuclear waste repository in crystalline host rock. *Journal of Hydrology*, *553*, 59–70.
76. Haitjema, H. M. (2005). *Analytic element modeling of groundwater flow*. Bloomington: Class-Pak Publishing.
77. Hajibeygi, H., Karvounis, D., & Jenny, P. (2011). A hierarchical fracture model for the iterative multiscale finite volume method. *Journal of Computational Physics*, *230*, 8729–8743.
78. Harbaugh, A. W., Banta, E. R., Hill, M. C., & McDonald, M. G. (2000). Modflow-2000, the U. S. geological survey modular ground-water model-user guide to modularization concepts and the ground-water flow process. *Open-file Report. U. S. Geological Survey*, *92*, 134.
79. Hartigan, J. A., & Manchek, A. W. (1979). Algorithm as 136: A k-means clustering algorithm. *Journal of the Royal Statistical Society. Series C (Applied Statistics)*, 100–108.
80. Hassanizadeh, S. M., & Leijnse, T. (1988). On the modeling of brine transport in porous media. *Water Resources Research*, *24*(3), 321–330.
81. Henry, H. R. (1959). Salt intrusion into fresh-water aquifers. *Journal of Geophysical Research*, *64*(11), 1911–1919.
82. Heroux, M. A., Bartlett, R. A., Howle, V. E., Hoekstra, R. J., Hu, J. J., Kolda, T. G., et al. (2005). An overview of the Trilinos project. *ACM Transactions on Mathematical Software (TOMS)*, *31*(3), 397–423.
83. Hidalgo, J. J., Carrera, J., & Medina, A. (2009). Role of salt sources in density-dependent flow. *Water resources research*, *45*(5),
84. Hughes, T., Mallet, M., & Mizukami, A. (1986). A new finite element formulation for computational fluid dynamics. II. Beyond SUPG. *Computer Methods in Applied Mechanics and Engineering*, *54*(3), 341–55.
85. John, V., & Knobloch, P. (2007). On spurious oscillations at layers diminishing (SOLD) methods for convection-diffusion equations: Part I - a review. *Computer Methods In Applied Mechanics And Engineering*, *196*(17–20), 2197–2215.

86. Joyce, S., Hartley, L., Applegate, D., Hoek, J., & Jackson, P. (2014). Multi-scale groundwater flow modeling during temperate climate conditions for the safety assessment of the proposed high-level nuclear waste repository site at Forsmark, Sweden. *Hydrogeology Journal*, *22*(6), 1233–1249.
87. Kaporin, I. (2007). Scaling, reordering, and diagonal pivoting in ILU preconditionings. *Russian Journal of Numerical Analysis and Mathematical Modelling*, *22*(4), 341–375.
88. Kaporin, I. E. (1998). High quality preconditioning of a general symmetric positive definite matrix based on its $U^T U + U^T R + R^T U$-decomposition. *Numerical Linear Algebra with Applications*, *5*(6), 483–509.
89. Kaporin, I. E., & Konshin, I. N. (2002). A parallel block overlap preconditioning with inexact submatrix inversion for linear elasticity problems. *Numerical linear algebra with applications*, *9*(2), 141–162.
90. Kapyrin, I., Ivanov, V., Kopytov, G., & Utkin, S. (2015). Integral code GeRa for raw disposal safety validation (in russian). *Gornyi zhurnal*, *10*, 44–50.
91. Kapyrin, I., Suskin, V., Rastorguev, A., & Nikitin, K. (2017). Verification of unsaturated flow and transport in vadose zone models using the computational code GeRa. *Voprosy Atomnoi Nauki i Tekhniki. Series: Mathematical Model in Physics Process*, (1):60–75.
92. Karypis, G., & Kumar, V. (1995). Metis–unstructured graph partitioning and sparse matrix ordering system, version 2.0. http://glaros.dtc.umn.edu/gkhome/metis/metis/overview.
93. Karypis, G., Schloegel, K., & Kumar, V. (2003). Parmetis. *Parallel graph partitioning and sparse matrix ordering library. Version, 2*. http://glaros.dtc.umn.edu/gkhome/metis/parmetis/overview.
94. Keyes, D. E., McInnes, L. C., Woodward, C., Gropp, W., Myra, E., Pernice, M., et al. (2013). Multiphysics simulations: Challenges and opportunities. *The International Journal of High Performance Computing Applications*, *27*(1), 4–83.
95. Kolditz, O., Ratke, R., Diersch, H.-J. G., & Zielke, W. (1998). Coupled groundwater flow and transport: 1. Verification of variable density flow and transport models. *Advances in Water Resources*, *21*(1), 27–46.
96. Konshin, I., & Kapyrin, I. (2017). Scalable computations of GeRa code on the base of software platform INMOST. In *International Conference on Parallel Computing Technologies* (pp. 433–445). Springer.
97. Konshin, I. N., Olshanskii, M. A., & Vassilevski, Y. V. (2015). Ilu preconditioners for nonsymmetric saddle-point matrices with application to the incompressible Navier-Stokes equations. *SIAM Journal on Scientific Computing*, *37*(5), A2171–A2197.
98. Kramarenko, V., Nikitin, K., & Vassilevski, Y. (2017). A finite volume scheme with improved well modeling in subsurface flow simulation. *Computational Geosciences*, *21*(5–6), 1023–1033.
99. Lacroix, S., Vassilevski, Y., Wheeler, J., & Wheeler, M. (2003). Iterative solution methods for modeling multiphase flow in porous media fully implicitly. *SIAM Journal on Scientific Computing*, *25*(3), 905–926.
100. Ladyzhenskaya, O. A., & Ural'tseva, N. N. (1968). *Linear and quasilinear elleptic equations*. Cambridge: Academic Press.
101. Langmuir, I. (1915). Chemical reactions at low pressures. *Journal of the American Chemical Society*, *37*(5), 1139–1167.
102. Lei, Q., Latham, J.-P., & Tsang, C.-F. (2017). The use of discrete fracture networks for modelling coupled geomechanical and hydrological behaviour of fractured rocks. *Computers and Geotechnics*, *85*, 151–176.
103. Li, L., & Lee, S. H. (2008). Efficient field-scale simulation of black oil in a naturally fractured reservoir through discrete fracture networks and homogenized media. *SPE Reservoir Evaluation & Engineering*, *11*, 750–758.
104. Li, N., Saad, Y., & Chow, E. (2003). Crout versions of ILU for general sparse matrices. *SIAM Journal on Scientific Computing*, *25*(2), 716–728.
105. Li, X. S. (2005). An overview of SuperLU: Algorithms, implementation, and user interface. *ACM Transactions on Mathematical Software (TOMS)*, *31*(3), 302–325.

106. Lipnikov, K., Svyatskiy, D., & Vassilevski, Y. (2009). Interpolation-free monotone finite volume method for diffusion equations on polygonal meshes. *Journal of Computational Physics*, *228*(3), 703–716.
107. Lipnikov, K., Svyatskiy, D., & Vassilevski, Y. (2010). A monotone finite volume method for advection-diffusion equations on unstructured polygonal meshes. *Journal of Computational Physics*, *229*, 4017–4032.
108. Lipnikov, K., Svyatskiy, D., & Vassilevski, Y. (2012). Minimal stencil finite volume scheme with the discrete maximum principle. *Russian Journal of Numerical Analysis and Mathematical Modelling*, *27*(4), 369–385.
109. Lipnikov, K., Svyatskiy, D., & Vassilevski, Y. (2013). Anderson acceleration for nonlinear finite volume scheme for advection-diffusion problems. *SIAM Journal on Scientific Computing*, *35*(2), A1120–A1136.
110. Malkovsky, V., & Pek, A. (2013). Effect of natural advection on stabilization of contaminant plume in natural traps at underground disposal of liquid wastes. *Water Resources*, *40*(7), 716–722.
111. Malkovsky, V., Yudintsev, S., Sharaputa, M., & Chulkov, N. (2019). Influence of buoyancy forces on movement of liquid radioactive waste from deep injection disposal site in the Tomsk region, Russian federation: Analytical estimate and numerical modeling. *Environmental Earth Sciences*, *78*(6), 219.
112. Moinfar, A., Varavei, A., Sepehrnoori, K., & Johns, R. T. (2013). Development of a coupled dual continuum and discrete fracture model for the simulation of unconventional reservoirs. In *SPE Reservoir Simulation Symposium*.
113. Nikitin, K., Terekhov, K., & Vassilevski, Y. (2014). A monotone nonlinear finite volume method for diffusion equations and multiphase flows. *Computational Geosciences*, *18*(3), 311–324.
114. Nikitin, K., & Vassilevski, Y. (2010). A monotone nonlinear finite volume method for advection-diffusion equations on unstructured polyhedral meshes in 3D. *Russian Journal of Numerical Analysis and Mathematical Modelling*, *25*(4), 335–358.
115. Nikitin, K. D., & Yanbarisov, R. M. (2020). Monotone embedded discrete fractures method for flows in porous media. *Journal of Computational and Applied Mathematics*, *364*, 112353.
116. Niswonger, R. G., Panday, S., & Ibaraki, M. (2011). MODFLOW-NWT, a Newton formulation for MODFLOW-2005. *US Geological Survey Techniques and Methods*, *6*(A37), 44.
117. Oldenburg, C. M., & Pruess, K. (1993). On numerical modeling of capillary barriers. *Water Resources Research*, *29*(4), 1045–1056.
118. Oldenburg, C. M., & Pruess, K. (1995). Dispersive transport dynamics in a strongly coupled groundwater-brine flow system. *Water Resources Research*, *31*(2), 289–302.
119. Olschowka, M., & Neumaier, A. (1996). A new pivoting strategy for gaussian elimination. *Linear Algebra and its Applications*, *240*, 131–151.
120. Olshanskii, M. A., Terekhov, K. M., & Vassilevski, Y. V. (2013). An octree-based solver for the incompressible Navier-Stokes equations with enhanced stability and low dissipation. *Computers & Fluids*, *84*, 231–246.
121. Olshanskii, M. A., & Tyrtyshnikov, E. E. (2014). *Iterative methods for linear systems: Theory and applications*. Philadelphia: SIAM.
122. Parkhurst, D. L., & Wissmeier, L. (2015). PhreeqcRM: A reaction module for transport simulators based on the geochemical model Phreeqc. *Advances in Water Resources*, *83*, 176–189.
123. Peaceman, D. W. (1978). Interpretation of well-block pressures in numerical reservoir simulation. *SPEJ*, *18*(3), 183–194.
124. Polubarinova-Koch, P. I. (2015). *Theory of ground water movement*. Princeton: Princeton University Press.
125. Richards, L. A. (1931). Capillary conduction of liquids through porous mediums. *Physics*, *1*(5), 318–333.
126. Ross, B. (1990). The diversion capacity of capillary barriers. *Water Resources Research*, *26*(10), 2625–2629.

127. Rybalćhenko, A., Pimenov, M., Kostin, P., Balukova, V., Nosuchhin, A., Mikerin, E., et al. (1998). *Deep injection of liquid radioactive waste in Russia*. Richland: Battelle Press.
128. Saad, Y. (2003). *Iterative methods for sparse linear systems*. Philadelphia: SIAM.
129. Saaltink, M. W., Ayora, C., & Carrera, J. (1998). A mathematical formulation for reactive transport that eliminates mineral concentrations. *Water Resources Research, 34*(7), 1649–1656.
130. Schäfer, M., Turek, S., Durst, F., Krause, E., & Rannacher, R. (1996). Benchmark computations of laminar flow around a cylinder. *Flow simulation with high-performance computers II* (pp. 547–566). Berlin: Springer.
131. Schneider, M., Agélas, L., Enchéry, G., & Flemisch, B. (2017). Convergence of nonlinear finite volume schemes for heterogeneous anisotropic diffusion on general meshes. *Journal of Computational Physics, 351*, 80–107.
132. Schöberl, J. (1997). Netgen, an advancing front 2d/3d-mesh generator based on abstract rules. *Computing and Visualization in Science, 1*(1), 41–52.
133. Šimůnek, J., & van Genuchten, M. T., (2008). Modeling nonequilibrium flow and transport processes using Hydrus. *Vadose Zone Journal, 7*(2), 782–797.
134. Shen, F., Kastrup, C. J., Liu, Y., & Ismagilov, R. F. (2008). Threshold response of initiation of blood coagulation by tissue factor in patterned microfluidic capillaries is controlled by shear rate. *Arteriosclerosis, Thrombosis, and Vascular Biology, 28*(11), 2035–2041.
135. Sleijpen, G. L., & Fokkema, D. R. (1993). BiCGstab (l) for linear equations involving unsymmetric matrices with complex spectrum. *Electronic Transactions on Numerical Analysis, 1*(11), 2000.
136. Soules, G. W. (1991). The rate of convergence of Sinkhorn balancing. *Linear Algebra and Its Applications, 150*, 3–40.
137. Steefel, C., Appelo, C., Arora, B., Jacques, D., Kalbacher, T., Kolditz, O., et al. (2015). Reactive transport codes for subsurface environmental simulation. *Computational Geosciences, 19*(3), 445–478.
138. Sterck, H. D. (2012). A nonlinear GMRES optimization algorithm for canonical tensor decomposition. *SIAM Journal on Scientific Computing, 34*(3), A1351–A1379.
139. Stockmann, M., Schikora, J., Becker, D.-A., Flügge, J., Noseck, U., & Brendler, V. (2017). Smart Kd-values, their uncertainties and sensitivities-applying a new approach for realistic distribution coefficients in geochemical modeling of complex systems. *Chemosphere, 187*, 277–285.
140. Stoyan, G. (1986). On maximum principles for monotone matrices. *Linear Algebra and Its Applications, 78*, 147–161.
141. Tene, M., Bosma, S. B., Al Kobaisi, M. S.,& Hajibeygi, H., (2017). Projection-based embedded discrete fracture model (pEDFM). *Advances in Water Resources, 105*, 205–216.
142. Terekhov, K. (2020). Collocated finite-volume method for the incompressible Navier-Stokes. *To appear in Journal of Numerical Mathematics*.
143. Terekhov, K. M., Mallison, B. T., & Tchelepi, H. A. (2017). Cell-centered nonlinear finite-volume methods for the heterogeneous anisotropic diffusion problem. *Journal of Computational Physics, 330*, 245–267.
144. Terekhov, K. M., & Vassilevski, Y. V. (2019). Finite volume method for coupled subsurface flow problems, I: Darcy problem. *Journal of Computational Physics, 395*, 298–306.
145. Todd, M. J., & Yıldırım, E. A. (2007). On Khachiyan's algorithm for the computation of minimum-volume enclosing ellipsoids. *Discrete Applied Mathematics, 155*(13), 1731–1744.
146. Van Genuchten, M. T. (1980). A closed-form equation for predicting the hydraulic conductivity of unsaturated soils 1. *Soil Science Society of America Journal, 44*(5), 892–898.
147. Van Reeuwijk, M., Mathias, S. A., Simmons, C. T., & Ward, J. D. (2009). Insights from a pseudospectral approach to the Elder problem. *Water Resources Research, 45*(4),
148. Varga, R. S. (1962). *Iterative analysis*. Berlin: Springer.
149. Vassilevski, Y., Konshin, I., Kopytov, G., & Terekhov, K. (2013). *INMOST - Program platform and graphic environment for development of parallel numerical models on general meshes*. Moscow: Moscow University Publishing. (in Russian).

150. Voss, C. I., & Souza, W. R. (1987). Variable density flow and solute transport simulation of regional aquifers containing a narrow freshwater-saltwater transition zone. *Water Resources Research*, *23*(10), 1851–1866.
151. Webb, S. W. (1997). Generalization of Ross' tilted capillary barrier diversion formula for different two-phase characteristic curves. *Water Resources Research*, *33*(8), 1855–1859.
152. Woods, J. A., Teubner, M. D., Simmons, C. T., & Narayan, K. A. (2003). Numerical error in groundwater flow and solute transport simulation. *Water Resources Research*, *39*(6),
153. Wufsus, A., Macera, N., & Neeves, K. (2013). The hydraulic permeability of blood clots as a function of fibrin and platelet density. *Biophysical Journal*, *104*(8), 1812–1823.
154. Younes, A., Ackerer, P., & Mose, R. (1999). Modeling variable density flow and solute transport in porous medium: 2. Re-evaluation of the salt dome flow problem. *Transport in Porous Media*, *35*(3), 375–394.

Printed by Printforce, the Netherlands